Food and Agriculture Organization of the United Nations

UNECE

ECE/TIM/SP/56

Developing National Forest Information Systems

A Practical Guide

UNITED NATIONS
Geneva, 2024

COPYRIGHT AND DISCLAIMER

This work is co-published by the United Nations (UNECE) and the FAO.

This publication is issued in English and Russian.

ACKNOWLEDGEMENTS

This practical guide document was drafted as part of the project on "Strengthening cooperation and national capacities in selected UNECE countries for sustainable forest management", implemented by the Joint UNECE/FAO Forestry and Timber Section and financially supported by the Russian Federation.

The practical guide document was jointly drafted by the author and the secretariat, drawing on information from national focal points through a dedicated survey, as well as available literature and data.

The Joint UNECE/FAO Forestry and Timber Section would like to acknowledge the contributions of - Andrzej Talarczyk (Author) and Bastian Stahl (Reviewer).

The secretariat would like to also express its gratitude to experts from countries of the UNECE region, for their contributions and the sharing of their knowledge on forest information management in their countries.

KEY POINTS

Good, informed policymaking and management requires information, monitoring outcomes, reporting and adjusting tools.

Information systems are an indispensable part of modern forest governance.

The use of forest information systems (FIS) is a cutting-edge innovation which significantly advances forest information management and brings numerous benefits to policymakers, all forest sector-related stakeholders and society in general.

Forest sector information systems should be aligned to forest management strategies, institutional landscapes, as well as to organizational, financial and overall administration capabilities.

Prior analysis and strategic action are crucial for the successful development and implementation of forest information systems.

ECE/TIM/SP/56

United Nations Publication

Sales No. : E.24.II.E.5

ISBN: 978-92-1-003014-4

e-ISBN: 978-92-1-358727-0

ISSN: 1020-2269

eISSN: 2518-6450

Contents

LIST OF ABBREVIATIONS

BULiGL	Bureau for Forest Management and Geodesy (forest management planning organization), Poland
C&I	Criteria and indicators
CSS	Communication support systems
DGLP	Directorate General of State Forest Holding, Poland
DSS	Decision support systems
FAO	Food and Agriculture Organization of the United Nations
FDB	Forest Data Bank (forest data platform in Poland)
FI	Forest inspectorate (forest management entity), Poland
FIS	Forest information system(s)
GIS	Geographic information system(s)
HA	High availability
ISS	Information-sharing systems
IT	Information technology
KPI	Key performance indicator
LAN	Local area network
NFI	National forest inventory
NGO	Non-governmental organisation
RMS	Resource management systems
SaaS	Software as a service
SFH	State Forests Holding (Polish forest management organization)
SFM	Sustainable forest management
SMART	Specific, measurable, achievable, relevant and time-bound
SOP	Standard operating procedure
SWOT	Strengths, Weaknesses, Opportunities, Threats
UNECE	United Nations Economic Commission for Europe
VCA	Value chain analysis
WAN	Wide area network

EXECUTIVE SUMMARY

Forest information systems (FIS) play a crucial role in effective forest management and decision-making processes. They provide a comprehensive and integrated approach to collect, analyse and disseminate data and information related to forest ecosystems. By developing robust forest information systems, owners, managers and other stakeholders can gain valuable insights into forest resources, forest health and vitality, biodiversity and related socioeconomic aspects. FIS enable evidence-based decision-making, support policy making and forest governance. Likewise, FIS facilitate monitoring and reporting, promote sustainable forest practices and timely management responses.

This guide aims to equip users with the knowledge and practical tools necessary to carry out a pre-analysis for the efficient development and implementation of forest information systems at the national level. The different sections include specific exercises and learning objectives, offering step-by-step instructions to help practitioners understand, develop and implement the outlined processes.

The exercises covered in the guide include institutional mapping, understanding the organizational context, as well as defining business and IT-related objectives of a FIS programme. Each exercise is accompanied by an introduction to the topic and explanations of the context.

The guide emphasizes the importance of active stakeholder engagement and participatory analysis throughout the FIS development process. It encourages users to identify key stakeholders, gather relevant information on the national context and use this data for informed decisions. Additionally, the guide provides guidance on mapping activities in the national forest sector, as well as establishing linkages between actors, activities and demands.

The wide range of services provided by forests is also reflected in the diversity of stakeholders. The composition and landscape of stakeholders varies from region to region and is subject to constant change. Therefore, any stakeholder analysis presented in this guide should only be considered as a snapshot of the situation on the ground. This applies also to the changing relationships between stakeholders. The examples used in this document are therefore not a simple blueprint and should only serve as illustrations that can vary according to local conditions.

Finally, the guide highlights the significance of assessing available and required IT resources, the importance of defining key performance indicators and identifying strategic actions to achieve the desired outcomes in a FIS development process.

INTRODUCTION

INTRODUCTION

The guide was designed for national analysts, policymakers and forestry professionals looking for a deeper understanding of the needs, opportunities and constraints of developing/improving forest information systems for forest sector activities and sustainable forest management (SFM). This publication guides users through the process of performing the strategic analysis.

This guide also provided the main training material for participants in a capacity-building workshop held in October 2023 in Bishkek, Kyrgyzstan, for the countries of the Caucasus and Central Asia.

This practical guide includes several exercises covering all stages of the analysis. The exercises define the potential scope of a forest information system development programme in the forest sector[1] . Throughout this process, an attempt is made to specify possible objectives, stakeholders involved, methods and means, as well as challenges of future implementation.

The exercises do not focus on technical implementation details. The analysis is more of a communication exercise for high-level stakeholders to ensure support and offer new insights by juxtaposing different points of view and needs.

For this reason, potential users should ideally have a fairly broad understanding of the forest sector in their country/organization/company in terms of policy and operational management, as well as forest information management. Familiarity with the practical implementation of information technologies and opportunities is helpful.

This publication complements the *Guidelines for Developing Forest Information Systems* which was also prepared under the UNECE project "Strengthening cooperation and national capacities in selected UNECE countries for sustainable forest management".

1 This guide refers to the development of information systems of the forest sector; however, the methodology, steps and exercises proposed can also be applied by other institutions, organizations or companies, within or outside the forest sector.

GLOSSARY OF TERMS

Some of the terms used frequently in this guide before describing the details involved in the analysis.

Forest Information

Forest information refers to the comprehensive collection of data and knowledge on spatial, ecological, social and economic insights about forest resources and their ecosystems, used to inform decision making and sustainable forest management.

Forest information is gathered from various sources, including aerial imagery, remote sensing, on-ground surveys and data collected from forest management activities. Information and data are subsequently processed, analysed and organised into a structured database or system to facilitate decision-making of stakeholders and support sustainable forest management.

Forest information management

Information management is the practice of identifying, collecting, storing and maintaining data and other types of information which are needed in the context of a particular organization or objective. It involves the gathering, dissemination, archiving and permanent removal of information in all its forms. Information management covers the procedures and guidelines organizations adopt to manage and communicate information among different individuals, departments and stakeholders.

Forest information management covers all stages of processing forest-related information, like data collection, aggregation, analysis and storage, related national and international reporting, dissemination and communication. Purpose and types of data include e.g., state and trends of forests ecosystems and resources, health conditions of forests, forest products and services, their management, trade and use, related policy making, etc.

Forest information system

Information systems are integrated sets of components (technology, networking communication, people and processes) that allow users to collect, process, store, organize and distribute data for providing information, knowledge and digital products, as well as provide a feedback mechanism to meet an objective. They are work tools that enable users to perform certain periodic repeatable tasks.

Forest information system (FIS) is a generic term for information systems used in forest, wildlife and environmental management as well as policymaking. A FIS is a system that provides forest managers, owners, decision makers or other stakeholders with accurate and up-to-date information about the forest resources and forest-related activities to support effective decision-making.

FIS play a critical role in the collection, management and dissemination of forest-related information and thereby support sound management decisions in the forest sector, including its public and private forests, the timber industry, forest policy and other stakeholders.

FIS provide essential data and information on forest resources, including their extent, distribution and conditions, as well as the social and economic benefits they provide. This information can include data on forest cover, tree species, ecological processes, timber trade and other human activities, as well as maps, reports and other forms of visual representation.

FIS can be understood in two ways:

a) The overall processes, subjects and stakeholders related to the collection, analysis, management, circulation, use and dissemination of forest-related information. FIS exist whether they are as such identified and described or not. FIS can be a set of different tools, methods and databases used to collect, process, store, analyse and disseminate information related to forests.

b) Information technology systems to aid gathering information to support informed choices, monitoring results and refining policies. It can include hardware, software and network components, as well as institutional and human resources to support them.

An integrated FIS is a structured approach of system design that manages and processes different types and sources of forest information and data made accessible and usable for a wide range of stakeholders within a single environment.

IT capacities (assets and skills) in the context of forest information systems

In the context of FIS, IT capacities are the assets and skills that are needed to enable and sufficiently carry out collection, digitization, storage, processing and dissemination of forest-related data and information.

The assets typically include hardware and software components (such as servers, computers, databases, networking devices, applications), physical facilities (such as data centres, communication networks and power supply systems), broadband Internet supply as well as trained human resources with sufficient expertise to manage and maintain the IT systems.

The skills are knowledge of individual human resource assets (employees and contractors), institutional memory (the stored knowledge within the organization) and organizational capabilities to perform tasks related to building, maintaining and operating FIS.

STAGES OF THE ANALYSIS

Various strategic planning frameworks help to structure the work of identifying objectives and actions, maintaining control and disseminating results. They are all based on a layered approach, from the general to the specific (Figure 1).

First, **(1)** external conditions are considered, then **(2)** needs and opportunities for intervention are identified, followed by **(3)** establishment of key performance indicators (KPIs) and **(4)** performance of a cost-benefit- analysis and finally **(5)** formulation of individual projects and their execution.

FIGURE 1. **Stages of the analysis (general approach)**

The general strategy for building an information system can be designed and implemented at different levels. Therefore, the planning and implementation framework should be adapted accordingly. The common approach, however, can be defined (Paulo et al. 2012), as presented in Figure 2.

FIGURE 2. **Five phases of IT governance implementation approach (after Paulo et al., 2012, modified by A. Talarczyk)**

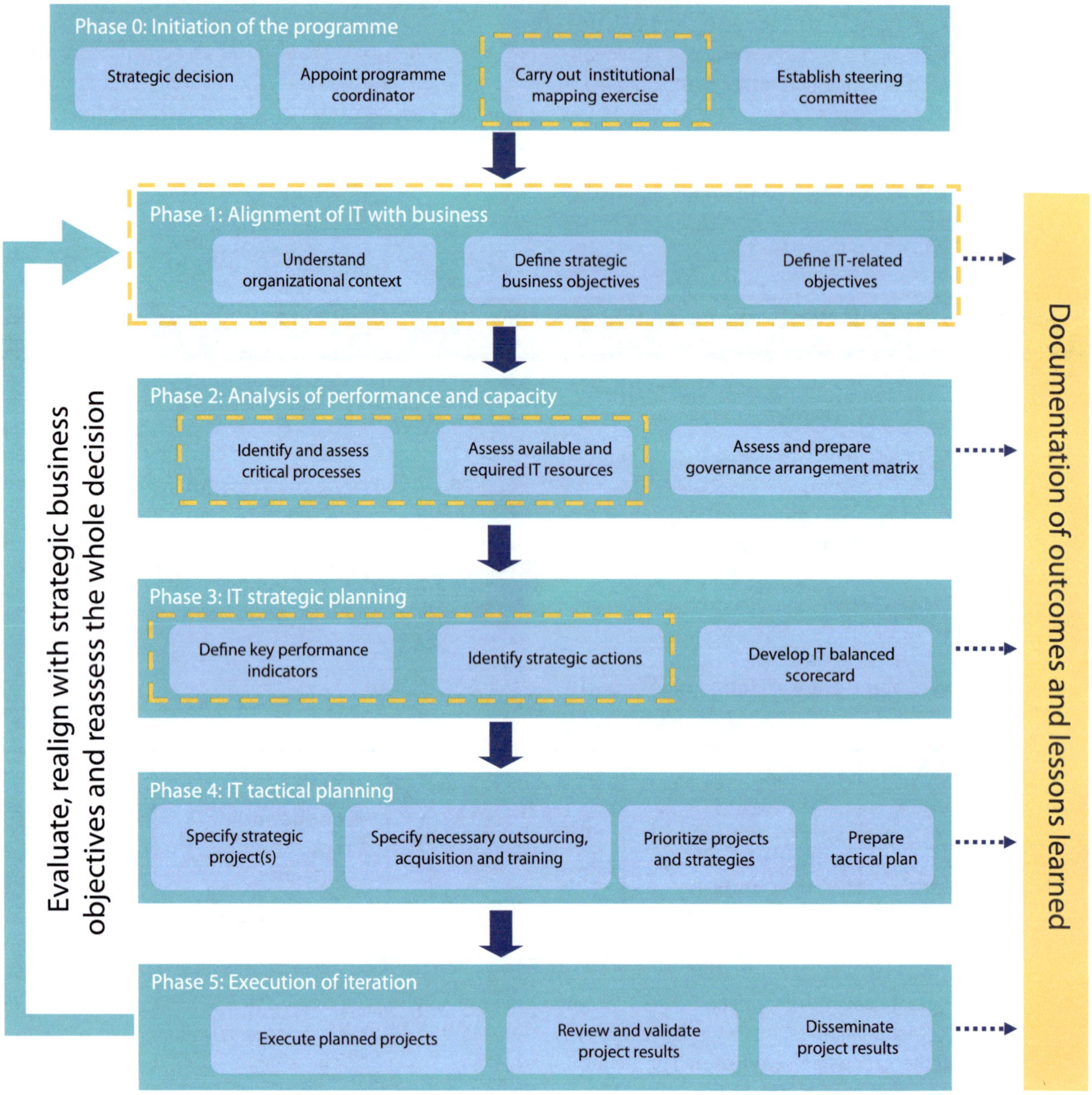

This guide concentrates on the following elements (framed in yellow in Figure 2) of a FIS's IT governance implementation approach:

- Institutional mapping;
- Understanding the organizational context (mapping crucial forest sector activities and identifying the most important stakeholders);
- Defining business objectives;
- Defining IT-related objectives;
- Identifying and assessing critical IT processes;
- Assessing available and required critical IT resources;
- Defining key performance indicators;
- Identifying strategic actions.

CO2
CO2
CO2
CO2

I. INSTITUTIONAL MAPPING

Objective of the exercise

Enhancing stakeholder understanding, promoting effective engagement in the context of forest information systems, managing stakeholder relationships and ultimately improving project outcomes.

Introduction

What is a stakeholder?

Stakeholders are organizations and social groups of any size or individuals that act at various levels and in different sectors (domestic, local, regional, national, international, private and public). These actors have a significant and specific stake in a given set of resources and can affect or be affected by resource management decisions.

What is a stakeholder analysis?

Stakeholder analysis is the process of systematically identifying, categorising, mapping and analysing information to determine:

- which groups have an **interest and/or influence** in a project;
- which groups are typically **included or excluded**;
- whether each group is **relevant to include**;
- which **concerns** are shared by various groups regarding the project;
- whether the groups **support or oppose** the project, or whether the groups will **benefit or be harmed** by it?

What needs to be analysed?

The process of identifying and analysing key stakeholders in a FIS analysis allows users to know their interests, positions, functions and roles, as well as their degree of involvement and ability or inability to affect policy change.

The mapping helps managers and organizations (e.g. government agencies or forest management enterprises) to gain a better understanding of who their key stakeholders are, their interests and concerns and how they can be effectively engaged or managed.

The mapping enables the identification of potential areas of collaboration or conflicts of interest, allowing for the development of appropriate strategies to address stakeholder needs and expectations. Stakeholder mapping also facilitates prioritizing stakeholder engagement efforts and ensuring that communication and decision-making processes are inclusive and transparent.

What are the steps in stakeholder mapping?

The following are the steps in stakeholder mapping:

1. Plan the analysis;
2. Identify stakeholders;
3. Gather information;
4. Fill in the Stakeholder Analysis Worksheet;
5. Analyse the Stakeholder Analysis Worksheet;
6. Apply the results.

STEP 1. PLAN THE ANALYSIS

Defining the purpose and scope of a stakeholder analysis is the crucial aspect of this preliminary step before the actual investigation is undertaken. This involves:

- Identifying the objectives users wish to achieve;
- Determining which stakeholders need to be involved;
- Specifying the information required from each stakeholder;
- Identifying the intended information users;
- Determining how the information will be used to benefit the project.

STEP 2. IDENTIFY STAKEHOLDERS

The first step in stakeholder identification is to create a list of groups that are known to have an influence and/or interest in forest sector policies and activities. This list serves as a starting point to further expand and analyse the identified stakeholders. To gather a comprehensive understanding of the stakeholders involved or impacted by the forest sector, a literature review is necessary. This involves conducting online and print searches of various sources, such as legal documents, websites, scholarly articles, company profiles, media publications, government and non-governmental organization (NGO) reports. Through this process, a wide range of stakeholders can be identified and analysed.

At the beginning of the stakeholder identification process, it is important to create a comprehensive list of all actors that may have a potential interest in forest information systems. This list should not be constrained by pre-existing knowledge and assumptions on their involvement in potential future projects involving FIS implementation.

During the analysis and stakeholder engagement stages, there will be an opportunity to confirm the relevance of each group's interest. Stakeholders can be categorized into the following groups (as outlined in the Stakeholder Analysis Worksheet, see Figure 22 in the annex): state entities, private sector, donors, civil society, international organizations, local NGOs, international NGOs, academia, local leaders or influencers and others.

> **Practical note:** The range of services provided by forests is also reflected in the diversity of stakeholders. The composition and landscape of stakeholders varies from region to region and is also subject to change. Therefore, any stakeholder analysis presented here should be considered as a snapshot of the real situation. This applies in particular to the relationships between stakeholders, which can also change. The examples used in this document are therefore not a simple blueprint but should only serve as illustrations that may vary according to local conditions.

STEP 3. GATHER INFORMATION

The initial lists generated through stakeholder identification rely on secondary sources. However, to gather more comprehensive information, it is important to reach out to project actors directly. There are three primary methods to collect information directly from stakeholders: (1) electronic communication, such as email or online surveys, (2) interviews conducted in person or via phone and (3) workshops of focus groups consisting of multiple stakeholders. These methods provide opportunities to gather more detailed and nuanced information on stakeholder interests, positions and concerns, as well as to establish stronger relationships with key stakeholders.

The goals of the analysis must be clarified before engaging stakeholders and be clearly communicated to them before conducting interviews. Goals help to identify and target those interests that need to be represented in the collaboration processes and those that can be left out considering that stakeholders will come to the interview with their own biases (Golder and Gawler 2005).

When conducting interviews, a standard questionnaire can be used based on the categories of the Stakeholder Analysis Worksheet. The specific type of interview questions should be tailored to the project or policy context, as well as the cultural context of the interviewed stakeholders. For example, questions may focus on stakeholders' positions on the issue at hand, inclusion or exclusion, alliances, conflicts, or other relevant topics.

To ensure consistency across interviews, it is recommended to develop a standard interview protocol that may include guidelines for how to ask questions, how to record the responses (e.g., notes or recording) and other relevant information. By using a consistent interview protocol, it will be easier to analyse and compare the responses from different stakeholders.

In addition to collecting information from known stakeholders, conducting interviews during the outreach phase can be a useful way to identify additional groups who may have an interest in, or be impacted by the project or policy. During the interviews, stakeholders can be asked to identify other groups who may not be on the initial stakeholder list but who may have relevant perspectives or concerns. This can help expand the list of stakeholders and ensure that all relevant groups are included in the stakeholder analysis process.

STEP 4. FILL IN THE STAKEHOLDER ANALYSIS WORKSHEET

The Stakeholder Analysis Worksheet should be filled in using the following guidance.

1. <u>What is the stakeholder's primary interest in FIS?</u>

Describe relations of stakeholder groups to forest information processing, their use and linked policies. For example, through expertise or knowledge of the area, as a key decision-maker, or by potential direct impact from developing FIS. Assign a value from 1-5 to the stakeholder's interest. If the stakeholder has a high degree of interest, a rating of 5 should be given, if a low degree, a 1.

2. <u>To what degree is the stakeholder group relevant to developing an FIS?</u>

Assign a value from 1-5, five being the highest relevance, for each group listed. Within this column, the following questions can be answered: What is the current role of the stakeholder?

What is the potential role of the stakeholder in the future FIS development? What is the potential contribution of the stakeholder to the FIS development?

3. To what degree is the group involved in the policymaking process?

A value from 1-5 should be assigned, 5 being the highest and a written explanation should be given if necessary – taking into account how this group has been involved in decision-making on policy and policy formulation both historically and presently.

4. Does the stakeholder group support or oppose the FIS development?

A spectrum of stakeholder types that can be provided is: supporter, moderate supporter, neutral, moderate opponent and opponent.

5. Will the developed FIS benefit or harm the stakeholder group?

Possible impacts are: benefit, harm, neither; how the groups will be affected should be explained.

6. What alliances exist with other stakeholders?

Any collaboration between relevant stakeholders should be listed and described.

7. What conflicts exist with other stakeholders?

Any conflicts between stakeholder groups should be described.

An example of the Stakeholder Analysis Worksheet content describing stakeholders of the forest sector in Poland is provided in Figure 23 in the annex.

Practical note: A large paper version of the Stakeholder Analysis template for workshop venue where participants add their comments in the table (see Figure 3) should be prepared. This will be useful for the analysis of the organizational context.

FIGURE 3. **A real-world example of filled Stakeholder Analysis Worksheet prepared during a workshop**

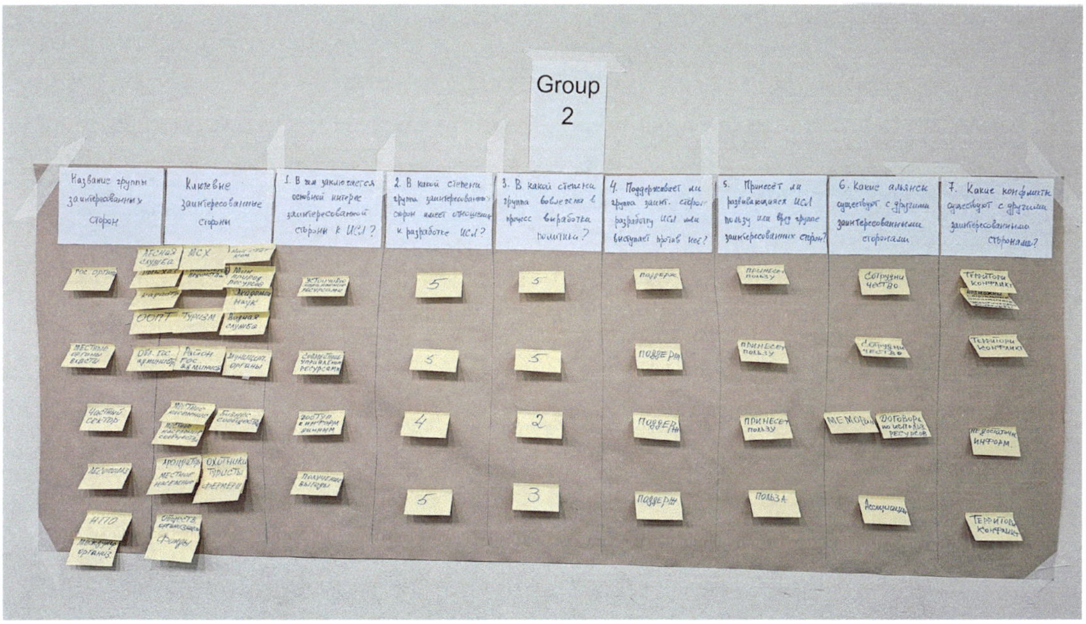

STEP 5. ANALYSE THE STAKEHOLDER ANALYSIS WORKSHEET

After gathering information on as many stakeholders as possible, the evaluator should analyse the data collected using the Stakeholder Analysis Worksheet. This analysis aims to gain a better understanding of each stakeholder group's interests, influence, positions, relevance, partnerships and conflicts.

The scale of 1 to 5 used in the Worksheet provides a simple way to assign a numerical value to the set of criteria, based on the evaluator's opinion of the stakeholder group's characteristics. However, opinions can vary. Therefore, it is recommended to seek information and feedback from as many actors as possible to ensure a more accurate rating.

In addition, the question *What is the stakeholder's primary interest in FIS?* is used to gain a clear understanding about the principal areas of interest of each stakeholder (or group of stakeholders). Furthermore, the response to this question helps to get to know which parts of the value chain they are linked to and what might be their likely position toward changes in the value chain elements.

The purpose of the question about degree of stakeholder relevance is to analyse how important particular stakeholders are for developing a FIS. An important issue is to realize what it means for the participants of this exercise to be "relevant" and to arrive at common understanding. This might not be so easy a task as opinions may be influenced by social factors and individual interests.

To what degree is the group involved in the policymaking process? The answer to this question depends on the formal structure of governance, but also on the actual estimated influence in decision-making.

Each group should be assessed to understand whether they support or oppose the project or policy to build FIS. In addition to classifying them into supporters, moderate supporters, neutral, moderate opponents, or opponents, it is important to look for the reasons for their position. It might happen that a group will be internally divided in their position, this should be described and, if possible, explained.

In addition to the perceived influence and position towards the project or policy, further analysis is recommended with a focus on analysing whether the stakeholder group will benefit or be harmed by the project. In particular, real or potential harm as well as financial or other benefits, should be analysed.

The final issue is understanding which alliances exist with other stakeholders. This form of linkage may foster collaboration and build support for the project or policy, but alliances might also be formed to oppose it. Therefore, if conflicts with other stakeholders exist (or could potentially appear) they need to be described.

STEP 6. APPLY THE RESULTS

The analysis should be documented in the form of diagrams with attached explanations/descriptions. One of the diagrams frequently used for this purpose (Golder and Gawler 2005) is shown in Figure 4.

FIGURE 4. **A diagram to summarize the stakeholder analysis (after Golder and Gawler 2005, modified)**

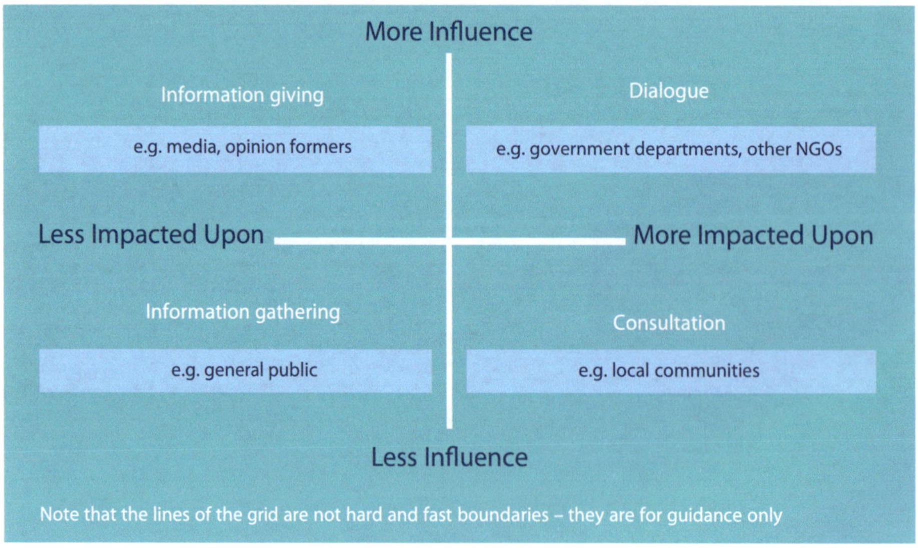

Another approach is to classify stakeholders by their power (influence) and interest. One may also visualize influence vectors – how the stakeholders can influence each other – see an example (Bryson 2018) in Figure 5.

FIGURE 5. **Stakeholder analysis with influence lines (after Bryson 2018)**

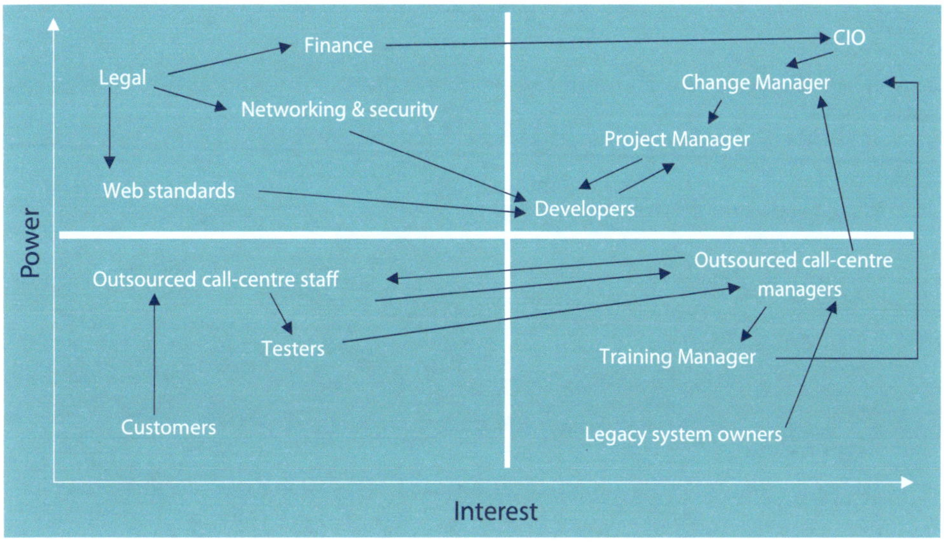

The four quadrants of such a matrix can be mapped to different communication strategies. An example of this approach (Storey et al. 2019) is shown in Figure 6.

FIGURE 6. **Stakeholder mapping used to tailor communication strategies (after Storey et al. 2019)**

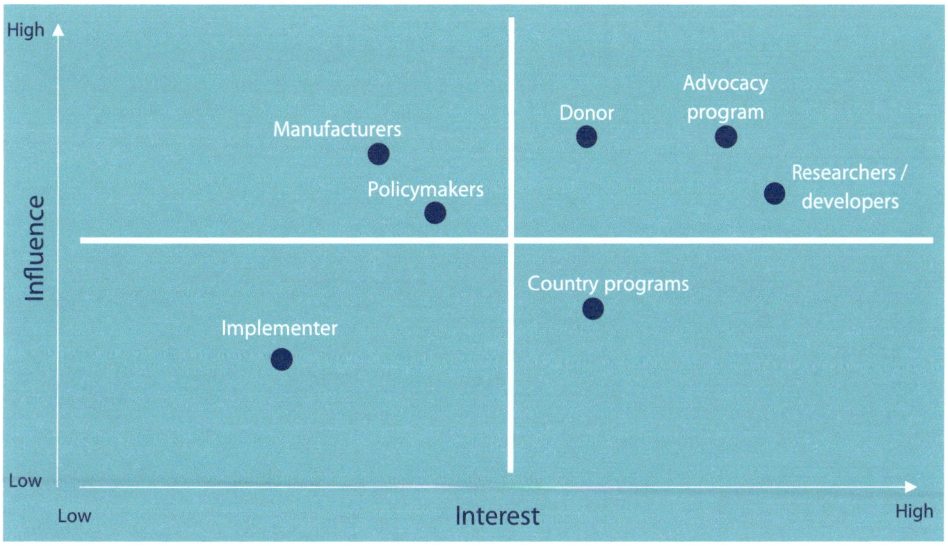

Practical note: The names of the stakeholders that have been identified earlier should be now placed on a diagram similar to those described in this section. A large sheet of paper with four named quadrants should be made available in the workshop venue. If sticky notes have been used previously, they can now be easily transferred to the analytical diagram. If necessary, vectors of influence can be then drawn directly on the diagram using table markers (see Figure 7).

FIGURE 7. **A real-world example of a diagram summarizing the stakeholder analysis prepared during a workshop**

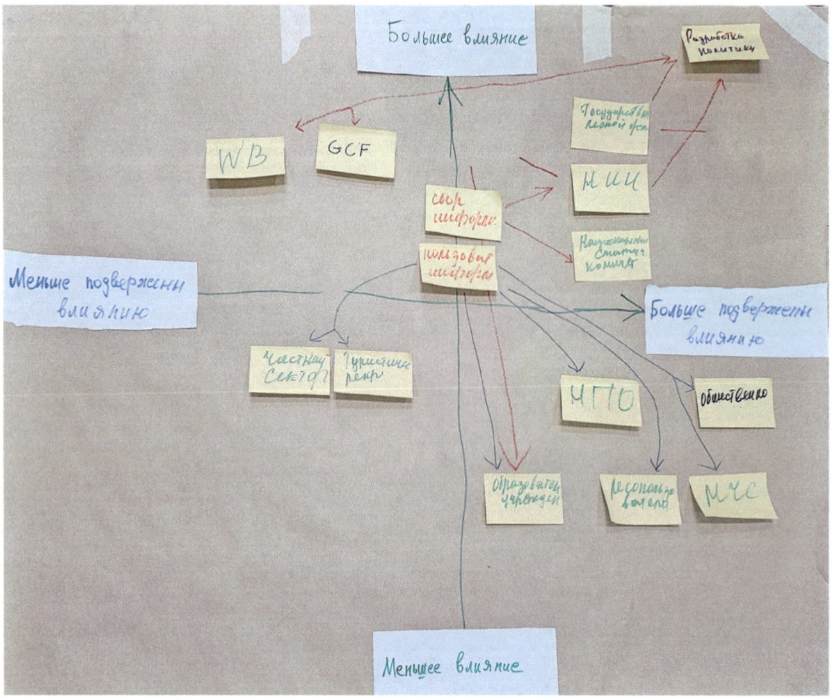

II. UNDERSTANDING THE ORGANIZATIONAL CONTEXT

Objective of the exercise

The objective of this exercise is to analyse how the forest sector is organized, what linkages and dependencies exist between various stakeholders and how they contribute to creating added value in the forest sector. These outputs will be used later for determining needs and opportunities for improvement through the development of an FIS.

Introduction

What is a value chain analysis?

Value chain analysis (VCA) is a frequently employed technique, especially in large-scale programmes, to develop an understanding of how activities and actors that are involved in bringing a product from production to consumption are linked. A value chain is understood here as the set of activities through which actors get increased value from products or services. "There is a simple element at the heart of value chain analysis. The idea of a chain is a metaphor for connectivity. It highlights the point that most goods and services are produced by a complex and sequenced set of activities" (Humphrey and Navas-Alemán 2010).

Why is value chain analysis important in FIS development?

A FIS is always embedded in a larger network of activities, which create relationships and constraints that have to be addressed and opportunities that can be exploited. Mapping of the value chain in which the FIS will be placed helps to identify (i) existing components, (ii) actors, (iii) activities and (iv) linkages – both vertical and horizontal and thus facilitates later a structured discussion about these opportunities, constraints and possible optimal solutions(Stein and Barron 2017).

What methods are used in value chain analysis?

The range of methods that are commonly used for researching value chains are:

- Desk studies;
- Interviews;
- Focus group discussions;
- Surveys;
- Document analysis; and
- Participatory observations.

As there is not a single, universally applicable method for studying value chains, usually a combination of them is used. However, the analysis should focus on the organizational context, not on details of particular elements of the value chain structure and/or their performance.

What are the steps of organizational context analysis?

The following are the steps of organizational context analysis:

1. Identify activities in the forest sector and business objectives behind them,
2. Identify actors related to the forest sector activities,
3. Establish linkages between the actors and relate the actors to the activities,
4. Add contextual notes.

If the scope of the digital transformation programme is limited to a single part of the value chain or a single organization, the organizational context analysis may be simplified.

FIGURE 8. **An outline of the value chain analysis**

	Steps	Task	Questions
		Creating a value chain map	
1	Activities	Identify activities of the forest sector and place them on the map. Identify and write down the business objectives that are behind each of these activities.	What activities are carried out in the forest sector which are necessary for its proper functioning? What are business objectives of each of these activities?
2	Actors	Identify actors related to the forest sector activities	Who are the key actors involved in the forest sector?
3	Links	Draw linkages between actors and relate the actors to activities	What links are important to understand how the forest sector operates and is governed?
4	Context	Write contextual factors on sticky notes and place them o the map	What factors outside the forest sector are important to consider?

STEP 1. IDENTIFY ACTIVITIES IN THE FOREST SECTOR AND BUSINESS OBJECTIVES BEHIND THEM

The exercise starts with brainstorming on the type of activities carried out in the forest sector (or part of it, depending on the analysis scope) which are necessary for its functioning. Create a list of these activities.

> **Practical note:** Names of activities should be written on sticky notes and placed on a whiteboard.

While listing activities, business objectives should be assigned to each activity. Business objectives are the measurable results that are aimed to be achieved by carrying out a specific activity in order to accomplish a longer-term vision.

> **Practical note:** Business objectives should be written on sticky notes and placed next to the appropriate activities.

STEP 2. IDENTIFY ACTORS RELATED TO THE FOREST SECTOR ACTIVITIES

Using the results of the institutional mapping exercise, key actors involved in the forest sector should be identified. It is likely that additional activities and business objectives will appear. In this case, the results of step 1 should be updated, too.

> **Practical note:** Names of actors should be written on sticky notes and placed next to the appropriate activities/business objectives.

STEP 3. ESTABLISH LINKAGES BETWEEN THE ACTORS AND RELATE THE ACTORS TO THE ACTIVITIES.

Once necessary objects have been identified, their linkages and dependencies should be analysed. The focus should be on those links that are important to understand how the forest sector operates and is governed. Those of marginal importance should be omitted. It is important to add descriptions of how and why the two objects are connected.

From a practical point of view, it is usually beneficial to present the linkages between the institutional stakeholders in the form of a graph (see Figures 10-12 as examples). The actual approach should be aligned with the content and objectives of the analysis.

> **Practical note:** The sticky notes that have been created in the previous steps should be clustered around related ones, leaving enough space for drawing links and placing other notes. Then appropriate links should be drawn. In case of mistakes or need to improve the object placement, the link should be erased and the analysis started again.

FIGURE 9. **Examples of stakeholder maps from various sectors**

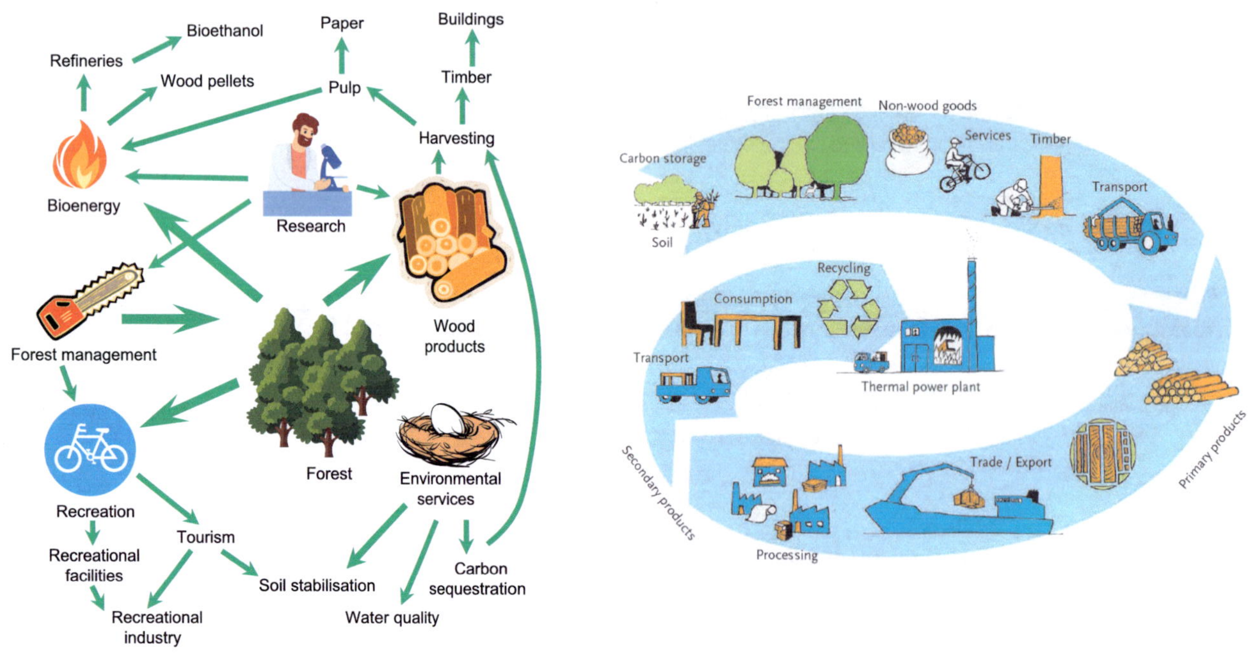

(Cavana and Adams, 2010)

(Wolfslehner et al., 2016)

It is beneficial to create two separate maps/graphs, presenting:

1. connections between stakeholders (see Figure 10) and
2. a network diagram of perceived stakeholder impact on other stakeholders (see Figure 11) (Svadlenak-Gomez et al. 2014).

FIGURE 10. **An example of network diagram of connections between stakeholders (after Svadlenak-Gomez et al., 2014)**

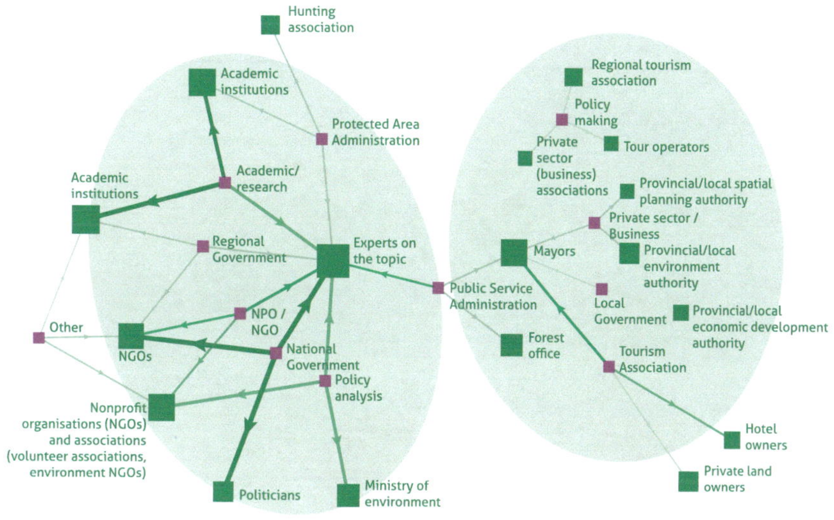

FIGURE 11. **An example of a network diagram of perceived stakeholder impact (after Svadlenak-Gomez et al., 2014)**

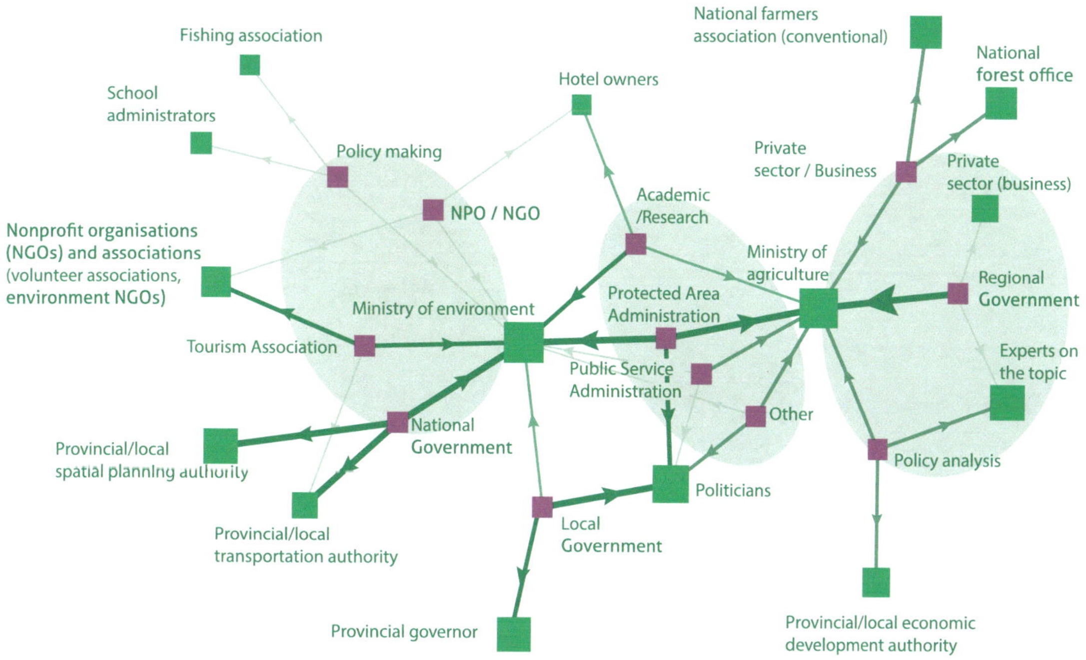

An alternative approach could be to visualize both linkages and influences in a single diagram on Figure 12 (Suoheimo et al. 2017).

FIGURE 12. **An integrated stakeholder map, including stakeholder connections, linkages and influences (after Suoheimo et al., 2017)**

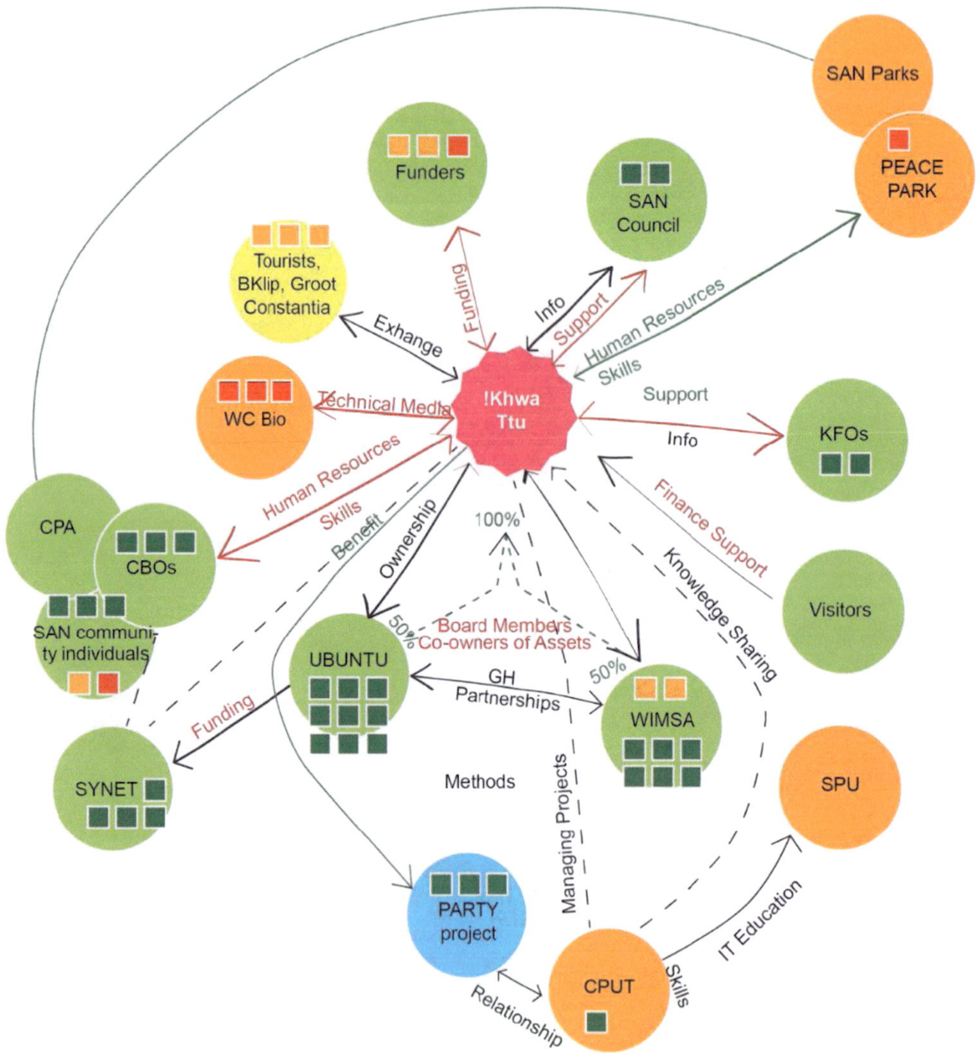

An example of stakeholder mapping of the forest sector in Poland is presented in Figure 13.

FIGURE 13. **An example of stakeholder mapping of the forest sector for Poland**

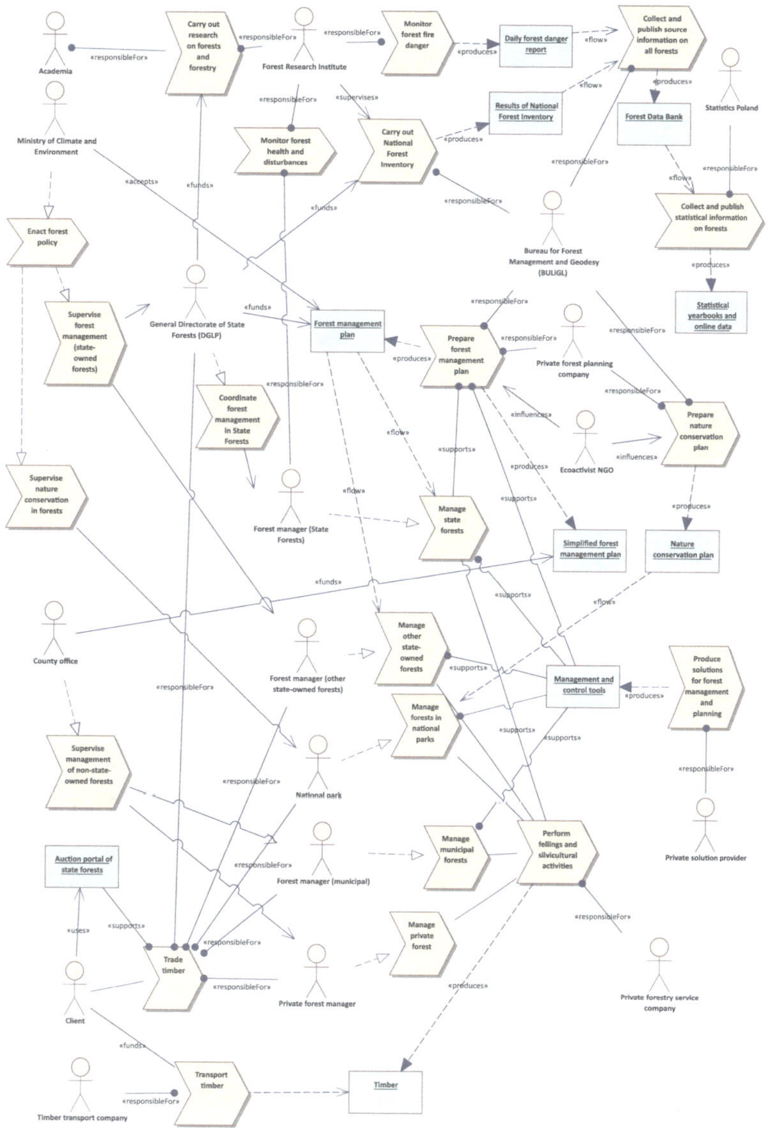

STEP 4. ADD CONTEXTUAL NOTES

In the context of developing and implementing forest information systems, adding contextual notes is a crucial step in the organizational context analysis. Contextual notes provide relevant information and insights about the actors, activities and linkages identified in the previous steps. These notes help in understanding the broader context within which the forest sector operates, enabling better decision-making and effective system design.

Practical note: To add contextual notes effectively, it is necessary to conduct interviews, surveys and research to gather comprehensive information about the forest sector and the actors involved. This should include socioeconomic factors, environmental considerations, regulatory frameworks, technological advancements and any other relevant contextual details. Additionally, perspectives and expectations of stakeholders should be considered to gain a well-rounded understanding.

III. DEFINING BUSINESS OBJECTIVES OF A FOREST INFORMATION SYSTEM PROGRAMME

Objective of the exercise

The objective of this exercise is to enhance identification and understanding of the key business objectives and goals that should be achieved through the implementation of FIS and understanding how FIS can support decision-making and value creation within the forest sector (or a part of the sector, depending on the analysis scope).

Introduction

What are the business objectives of a FIS programme?

The business objectives of a FIS development programme/ strategy are fundamentally different from the forest sector business objectives and one should take caution not to mix the two notions. While the forest sector business objectives are the motivation for activities of the forest sector stakeholders, the FIS programme business objectives describe how successful execution of the FIS development programme should influence the forest sector (or its part).

For example, assuming that a legally established duty of a forestry government agency is communication with citizens on forest-related issues, *the agency's business objective* would be ***providing comprehensive information on the country's forest resources to the public***. This business objective would be realized through the activity of sharing data on forests over the Internet. While analysing possible points of intervention through a FIS, one of *the business objectives of the FIS programme* could be ***improving the quality and scope of the agency's communication with the public***.

What is the purpose of defining business objectives of the FIS programme?

The goal of value chain analysis and value chain mapping is to better understand the current system in order to identify strategies for improving. Business objectives of the FIS programme depend to a large degree on policy objectives, but it is vital to identify the most important business objectives of the FIS programme through a structured discussion informed by the VCA results and input from members of the analytical team (steering committee). Resources available for developing the FIS are usually limited and care should be taken to identify and choose the FIS programme business objective so that maximum cost/outcome value is achieved.

How should points of intervention for a FIS programme be identified?

There are many types of value chain interventions that lead to defining concrete business objectives. Interventions (see Figure 14) can be classified for example as the following four types (Humphrey and Navas-Alemán 2010):

1. **Strengthening the weakest link.** This type of intervention addresses the issue of weak links in the chain. The efficiency of the chain as a whole depends upon the performance of each of the linkages. A value chain intervention starts from a diagnosis of the impact of a particular link in the chain on the performance of the chain as a whole.
2. **Improving knowledge and resource flows.** Poor flows of knowledge are commonly found in value chains. Knowledge is either poorly communicated, or not communicated at all. This leads to missed opportunities or to inefficiencies in the value chain operation. One implication of targeting these flows is that knowledge deficits at one point in the chain may be addressed by interventions involving actors at other points in the chain.
3. **Improving linkages.** The efficiency of a value chain depends on the readiness of the actors to exchange information and collaborate. Poor linkages, e.g. lack of trust between parties involved, result in high transaction costs or transaction failures. Value chain interventions address these problems in a variety of ways, including trust-building initiatives, providing indicators of organization/company reputation and setting up institutions to manage risk. While in practice improving linkages is closely related to the second objective, improving knowledge and resource flows, they are analytically distinct.
4. **Developing new or alternative links in the chain.** If the overall efficiency of the chain depends on the functioning of each of the links, value chain interventions can identify alternative linkages. Linkage A shows the development of new sources of supply to overcome problems such as shortage of raw materials, or low quality. Linkage B shows how one link in a chain is replaced by another. This might be the brokering of new intermediaries to substitute a link that was malfunctioning, or non-existent. Linkage C shows the case of new market outlets being developed. In each case, the new link might complement or replace the old one.

There is considerable underutilized potential in value chains that can be tapped into by improving flows of knowledge and resources, improving relationships and establishing linkages, as well as by improving capabilities at specific points in the chain. Outside interventions may be needed to unlock this potential. In other words, strategies to improve the forest sector functioning should focus as much on linkages as on its assets.

FIGURE 14. **Typology of objectives for value chain interventions (Humphrey and Navas-Alemán, 2010, modified)**

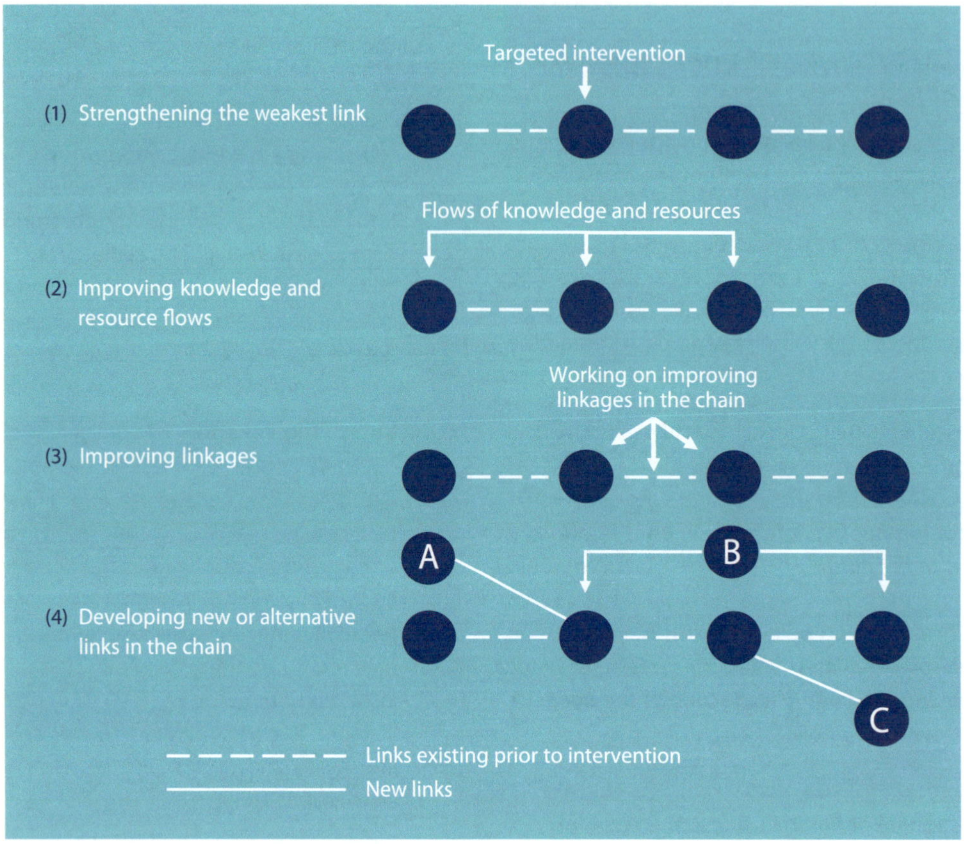

What are the steps in defining business objectives of an FIS development programme?

The steps in defining business objectives of an FIS development programme (Figure 15) are:

1. Participatory analysis;
2. Validation of the findings;
3. Identifying solutions.

FIGURE 15. **Defining business objectives and upgrading strategies**

	Steps	Task	Questions
		Analysing and interpreting the value chain map	
1	Participatory analysis	Identify possible points of intervention using the intervention typology and attempt to quantify expected benefits that would result from each intervention. Mark them on the map.	What activities are carried out in the forest sector which could be improved/changed to improve its functioning?
2	Validation	Summarize, discuss and ensure coherency of emerging findings.	Does the map represent the forest sector's reality adequately? Is the map coherent?
3	Identifying solutions	Identify and rank strategies for improvement. Ensure that each proposed strategy has clear business objectives.	What strategies are most effective to improve functioning of the forest sector?

STEP 1. PARTICIPATORY ANALYSIS

In this step, the task is to identify possible points of intervention using the intervention typology described in the introduction and attempt to quantify expected benefits that would result from each intervention. They should be marked on the value chain (context) map prepared in the previous exercise.

The question to be addressed is what activities carried out in the forest sector could be improved/changed using information technologies in order to have a positive impact on the sector.

> **Practical note:** This step could be organized as separate discussion groups taking notes of their proposals for intervention, together with justifications and marking them on the value chain map.

STEP 2. VALIDATION OF THE FINDINGS

After the brainstorming session on possible points of intervention, the findings should be summarized and discussed. Proposed items should be evaluated for their relevance, urgency, cost and expected benefits. Care must be taken to ensure coherence between all proposed business objectives of a FIS development programme.

STEP 3. IDENTIFYING SOLUTIONS

The final step is choosing from the proposed points of intervention those items which are deemed to be most suitable to the current situation. They should be ranked with business objectives assigned to each of them. There is no single approach to identifying and ranking solutions.

Different factors should be considered, including: political priorities, legal obligations, existing strategies and policies, cost-effectiveness, available resources (both financial and human), organizational circumstances etc.

For a FIS development programme more limited in scope, there may be only one or a few points of intervention with a limited number of business objectives.

All identified solutions should be listed and ranked with their respective business objectives to be achieved within the FIS development programme.

IV. DEFINING INFORMATION TECHNOLOGY-RELATED OBJECTIVES

Objective of the exercise

Business objectives defined in the context of the current policies and the forest value chain must be translated into actionable IT objectives. IT-related objectives establish clear targets and outcomes that guide the development, implementation and management of FIS, aligning them with the specific needs and goals in the context of the business objectives.

Introduction

In the context of forest information systems, defining IT-related objectives is a critical process that aligns IT projects with the overarching business objectives of the forest sector. This step ensures that the IT initiatives undertaken are purposeful and geared towards delivering specific outcomes that contribute to the achievement of broader goals. By clearly outlining the IT-related objectives, stakeholders can prioritize and plan IT projects effectively, optimizing resources and maximizing the impact of technological investments.

IT-related objectives serve as the bridge between the business objectives of the forest sector and the technical implementation of FIS. They define the specific technology-related targets and outcomes that need to be accomplished to support and enhance various activities within the sector. These objectives are framed in a manner that reflects the benefits that technology can bring to the sector's functioning and long-term growth.

By defining clear and measurable IT-related objectives within the framework of FIS, organizations can ensure that their technology projects are purpose-driven, aligned with business goals and yield tangible benefits for the forest sector and its stakeholders.

Some examples of possible IT-related objectives in the forestry sector:

- Enhanced forest monitoring and management through remote-sensing technology;
- Strengthened forest supply chain transparency with blockchain technology;
- Enabled data-driven decision-making with advanced analytics.

What are the steps in defining IT-related objectives of an FIS development programme?

The steps in defining IT-related objectives of an FIS development programme are:

1. Formulating IT-related objectives;
2. Validating IT-related objectives;
3. Selecting candidates for intervention (optional).

STEP 1. FORMULATION OF IT-RELATED OBJECTIVES

The starting point for this exercise is the list of business objectives of the FIS programme as outlined before. They represent the possible points of intervention in the value chain to improve its functioning. They should be translated into actionable IT-related objectives. An example list of IT-related objectives formulated in the course of initiating the Forest Data Bank development in Poland is listed in Figure 16.

Technical solutions are not expected to be defined at this stage because this would require a detailed process and resource analysis which will be done in Phase 2 (see Figure 2). At this point, the process should focus on identifying IT-related product or solution ideas that will support the achievement of the business objectives.

For example, if a business objective was defined as providing the general public with consistent, timely and accurate information on forest resources, the corresponding IT objective could be: to create a national repository to collect, store, process and publish data on forests.

Another business objective could be: to monitor and prevent illegal logging and deforestation. Its corresponding IT objectives would be: (1) to create a timber-tracking system, (2) to create an early-detection deforestation monitoring system.

FIGURE 16. **An example list of IT-related objectives formulated in the course of initiating the Forest Data Bank development in Poland**

	Business objective	IT-related objective	Description and justification (how this solution can contribute to achieving of the business objective)
1	Build a central repository of information on forests in Poland	Build a data warehouse and put in place procedures for forest-related data provision	There is no single place where all data on the Polish forests could be found. Supervision of private forests especially is decentralized; Collecting all forest inventory data in one place will create added value and significant synergies.
2	Improve communication of forest information to the public	Build a web portal for comprehensive presentation of information on Polish forests	There is no dedicated website where users could get comprehensive information on the Polish forests. Building a web portal will create new value for the forest managers and the public in terms of public communication.
3	Improve communication of forest information to the public	Build a general-purpose mobile app for comprehensive presentation of information on Polish forests	A general-use mobile app will increase knowledge on forests in Poland, will promote tourism and will increase social awareness of forest-related matters.

Practical note: This step could work well when participants work in separate discussion groups with each having the same value chain map and the same list of business objectives to work with, especially if there are more than five participants. Splitting into smaller groups prevents reaching the consensus prematurely and thus potentially silencing minority views with important insights. The groups should use sticky notes for assembling their ideas on IT-related products or solutions with a potential to help attain each business objective adding justifications. These could be placed on their copy of the value chain map, next to the respective business objective.

STEP 2. VALIDATION OF THE IT-RELATED OBJECTIVES

After completion of Step 1, findings should be summarized and discussed. Proposed items should be evaluated considering:

- Their relevance and alignment to the corresponding business objective, i.e., how much they can contribute to achieving the business objective;
- Effectiveness, i.e., whether they are the best imaginable solution for achieving the respective business objective; for example, if several solutions/products have been proposed, they should be evaluated and scored by their complexity, cost, etc. to identify the best one;
- Feasibility, i.e., how easy it would be to develop and maintain these particular solutions/products.

Care must be taken to ensure coherence between the proposed IT-related objectives for a FIS development programme.

Practical note: Having completed the work in subgroups, the teams convene and present their ideas. The moderator keeps a table of all proposed IT-related objectives, taking notes on their pros and cons for relevance, effectiveness and feasibility (see Figure 17).

STEP 3 (OPTIONAL). SELECTING CANDIDATES FOR INTERVENTION

If members of the group feel competent, they might attempt to choose IT-related objectives to address future FIS-building projects. The selection of IT-related objectives is not an easy process, as it requires attention to many factors, some of which are general (e.g. cost, availability of technology) but others are specific to the context of the overall analysis (e.g. policy priorities, available resources, organizational circumstances, governance framework).

FIGURE 17. **A real-world example of the table of business objectives and corresponding IT-related objectives developed during a workshop**

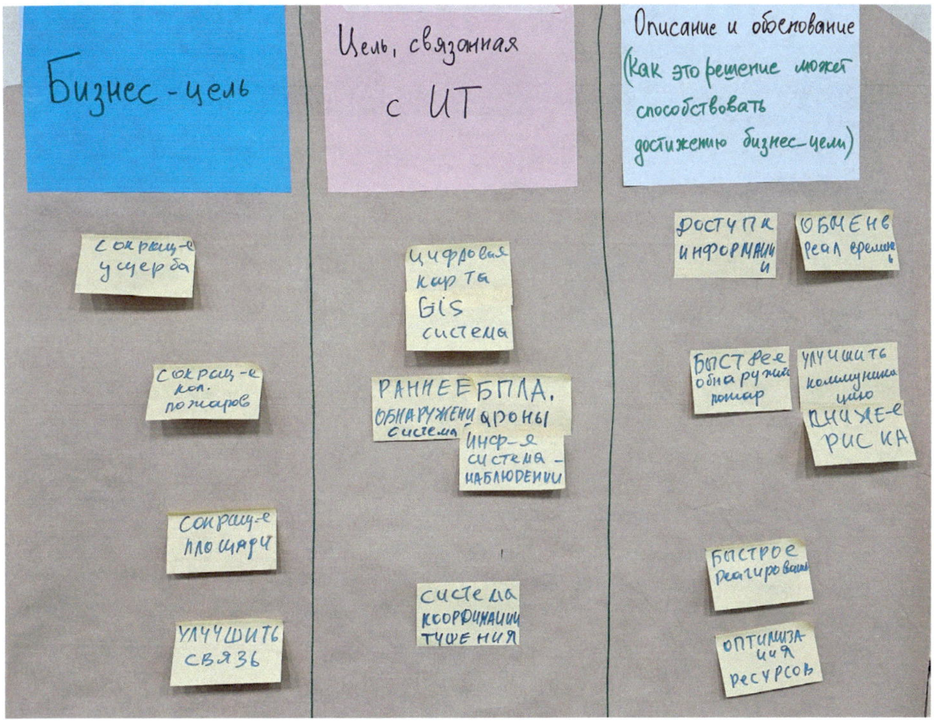

V. IDENTIFYING AND ASSESSING AVAILABLE AND REQUIRED CRITICAL IT PROCESSES

Objective of the exercise

The objective of the exercise is to identify the IT-related processes whose existence may be critical for achieving the business objectives and to assess their current level of maturity (or development) and completeness. The purpose of this activity is to gain additional insight into the possible areas of intervention for developing or improving the information system(s).

While the previous exercise was devoted to formulating the IT-related *objectives*, attainable by means of information technology, which would support the business objectives. This activity aims at assessing what IT-related *processes* might be necessary to put in place (or improve if they already exist) to facilitate reaching these objectives.

Introduction

Assessing the maturity level of IT processes from a business perspective is crucial for several reasons:

1. **Alignment with business goals**: A maturity assessment ensures that IT processes are aligned with the business's strategic goals. Mature IT processes are better equipped to support and drive business objectives, leading to improved overall performance and competitive advantage.
2. **Optimization of resources**: Mature IT processes often result in more efficient use of resources. This includes optimizing staff time, reducing wastage and making better use of technological investments, all of which can lead to cost savings and improved productivity.
3. **Risk management**: A higher maturity level in IT processes generally equates to better risk management. Mature processes have well-defined procedures for handling risks and contingencies, thus reducing the likelihood of disruptions to business operations.
4. **Quality of service**: Mature IT processes improve the quality of IT services and products. This results in higher customer satisfaction, increased reliability and better service delivery, which are all critical for maintaining a positive business reputation.
5. **Innovation and adaptability**: Businesses with mature IT processes are often more agile and innovative. They can adapt more quickly to changing market demands, new technologies, or disruptions, which is vital in today's fast-paced business environment.
6. **Regulatory compliance**: Many industries and administrative environments have stringent regulatory requirements regarding data security, privacy and operational standards. Mature IT processes help ensure compliance with these regulations, reducing the risk of legal issues and penalties.
7. **Decision-making and planning**: Mature IT processes provide better metrics and data for decision-making. This supports more accurate and strategic business planning budgeting and forecasting.
8. **Employee satisfaction and productivity**: Well-defined and efficient IT processes can lead to higher employee satisfaction, as employees have clear guidelines and tools to perform their tasks. This often translates into higher productivity and lower staff turnover.
9. **Market competitiveness**: Businesses with mature IT processes are often more competitive in the marketplace. They are able to leverage technology effectively, respond quickly to market changes and offer better products and services.
10. **Scalability and growth**: Mature IT processes are scalable, making it easier for organizations to grow and expand. Scalable processes accommodate increased workloads and organizational changes without a significant drop in performance or service quality.

In summary, assessing and improving the maturity level of IT processes is vital for ensuring that the IT function effectively supports and enhances business objectives, maintains competitiveness and adapts to the ever-evolving business and technological landscape.

What are the steps of identifying and assessing available and required critical IT processes?

The steps in identifying and assessing available and required critical IT processes are:

1. Determination of critical IT processes;
2. Assessment of current maturity level;
3. Performance of gap analysis.

STEP 1. DETERMINE CRITICAL IT PROCESSES

Identifying critical processes starts with determining which IT processes are critical for achieving the defined business objectives. This could include processes like software development, data management, IT security and network infrastructure management. This should be done in collaboration with stakeholders across the organization, including business unit leaders, end users and IT staff. Their insights will be valuable in determining which IT processes are essential for daily operations and strategic initiatives.

Then map the determined critical IT processes to business functions (activities). Analyse how various IT processes support specific business functions. This mapping helps in determining which IT processes are critical for the functioning of key business areas such as sales, operations, finance and customer service.

Evaluate the impact of each IT process on the business. Consider what would happen if a process failed or were interrupted. Processes that have a high impact on business continuity or those on which multiple other processes depend are typically critical.

Determine IT processes that are essential for managing risks and meeting compliance requirements. Processes that safeguard data security, privacy and regulatory compliance are often critical. Analyse historical data regarding IT incidents, downtime and performance bottlenecks. Processes that have previously caused significant disruptions or frequent bottlenecks in performance are likely critical.

Prioritize processes based on the value they deliver to the business and their efficiency. Critical processes are often those that add significant value or require optimization to improve overall business performance.

This determination of critical IT processes is the first step in ensuring that these processes are managed effectively, optimized and aligned with the overall business objectives.

> **Practical note:** This step could be organized as separate discussion groups taking notes of their proposals of identified critical processes on post-it cards, together with justifications and placing them on a large sheet of paper, one under another, together with the corresponding business objectives.

STEP 2. ASSESS CURRENT MATURITY LEVEL

Various maturity models, such as the Capability Maturity Model Integration (CMMI) or the IT Infrastructure Library (ITIL), are widely used to assess the current maturity level of IT processes (CMMI Product Team and others 2006; Gërvalla et al. 2018). This assessment highlights areas of strength, identifies gaps and helps in formulating a structured approach to elevate the maturity of IT services, ultimately ensuring they deliver maximum value to the business.

In this exercise, the ITIL (Information Technology Structure Library) framework is used. ITIL categorizes IT service management into several processes and functions, such as Service Strategy, Service Design, Service Transition, Service Operation and Continual Service Improvement. Each area consists of specific processes (since ITIL v4 the term 'practices' is also used which conveys a broader meaning as they can group several processes). An ITIL maturity assessment involves evaluating how well an organization has implemented these practices.

ITIL does not prescribe a specific maturity model but is often assessed using a simple scoring scale. This scale typically ranges from Level 1 (ad hoc activities) to Level 5 (highly organized and self-optimizing processes). The assessment involves determining how well an organization's IT processes and practices align with ITIL recommendations at each maturity level.

The capability levels of every process can be described as follows:

- **Level 1:** The process is not well organized; it is performed as initial/intuitive. It may occasionally or partially achieve its purpose through an incomplete set of activities.
- **Level 2:** The process systematically achieves its purpose through a basic set of activities supported by specialized resources.
- **Level 3:** The process is well defined and achieves its purpose in an organized way, using dedicated resources and relying on inputs from other processes that are integrated into a service management system.
- **Level 4:** The process achieves its purpose in a highly organized way and its performance is continually measured and assessed in the context of the service management system.
- **Level 5:** The process is continually improving organizational capabilities associated with its purpose. Each capability level is based on the previous ones; they need to be achieved before the current level can be assessed.

The assessment process begins with collecting data on the organization's IT-related processes and practices. This includes reviewing process documentation, conducting interviews with IT staff and stakeholders and analysing service performance data. The goal is to gain a comprehensive understanding of how IT services are managed and delivered. The full description of a comprehensive review process for an organization is beyond the scope of this manual. The reader is referred to the official ITIL Process documentation[2] or one of many books on the subject.

The collected data is then compared against ITIL best practices. This analysis helps identify areas where the organization's processes do not fully align with ITIL recommendations. It is important to consider both the efficiency and effectiveness of processes in delivering value to the business.

For each ITIL process or function, the organization's current practice (i.e., the way the process is organized and executed) is rated on the maturity scale. This involves assessing aspects like process standardization, documentation, integration with other processes, performance measurement and continuous improvement mechanisms.

Practical note: This step can be done as separate discussion groups taking notes of their evaluation of the maturity level of identified critical processes on post-it cards, together with justification of the assessment. After a group session, each group leader places the post-it cards next to the processes identified in the previous step. Then, an arithmetic average is calculated for each process. Identified areas for improvement can be noted on the side, they will help later identify strategic actions.

STEP 3. PERFORM GAP ANALYSIS

Compare the current maturity levels of the IT processes with their desired levels required to achieve the business objectives, then identify gaps and areas that need improvement.

2 https://www.itlibrary.org/

VI. ASSESSING AVAILABLE AND REQUIRED CRITICAL INFORMATION TECHNOLOGY RESOURCES

Objective of the exercise

The objective of the exercise is to conduct a Strengths, Weaknesses, Opportunities and Threats (SWOT) analysis of the available and required critical IT resources in the forestry sector's information processing domain to facilitate effective strategic project planning aligned with business goals.

Introduction

To effectively plan strategic projects leading to improvements defined in the business goals, it is necessary to identify how and where information technologies are employed (*processes*) and what *resources* (see the next chapter) currently exist in relation to information processing in the forestry sector. It is part of the effort to evaluate the performance and capacity of IT to meet the IT objectives defined in previous exercises.

The approach used is a SWOT analysis of each resource assessed and considered important for the process of FIS development.

The following types of IT resources should be included in the analysis:

- existing IT systems;
- qualified personnel;
- communication infrastructure;
- servers and other backend infrastructure;
- software-as-a-service (SaaS) solutions available; and
- available IT services.

Existing IT systems

Existing IT systems include all the software applications and hardware infrastructure that the forestry sector currently uses to process and manage information. It may involve data management systems, inventory tracking software, analytics tools, GIS (Geographic Information System) and more. These systems are critical as they form the backbone of information processing and decision-making within the organization

Qualified personnel

Qualified personnel refers to the skilled and knowledgeable workforce responsible for managing, maintaining and utilizing IT resources. Having qualified personnel is crucial to ensure the smooth functioning of IT systems, troubleshoot issues and optimize processes to derive valuable insights from the data.

Communication infrastructure

Communication infrastructure encompasses the network and communication channels that enable data transfer and collaboration. It includes local area networks (LANs), wide area networks (WANs), internet connectivity and other communication protocols. A robust communication infrastructure is vital for seamless data-sharing and real-time collaboration, especially in geographically dispersed forestry operations.

Servers and backend infrastructure

Servers and backend infrastructure are the physical or virtual servers that store and process data, along with the supporting infrastructure like data centres, storage systems and cloud platforms. Reliable servers and backend infrastructure are critical for data storage, processing and retrieval, ensuring data accessibility and security.

Software-as-a-Service solutions at disposal

Software-as-a-Service (SaaS) solutions are cloud-based software applications provided by third-party vendors. They offer flexibility and scalability without the need for extensive on-premises infrastructure. SaaS solutions used in the forestry sector might include data analytics platforms, project management tools and other specialized applications.

Available IT services

Available IT services refer to outsourced or managed services that the forestry sector can leverage, such as IT consulting, cybersecurity services, data analytics services and software development services. These services can supplement internal capabilities and expertise.

What are the steps in assessing available and required critical information technology resources?

The steps in the assessing available and required critical information technology resources are as follows:

1. SWOT analysis; and
2. Evaluation and identification of points of intervention.

STEP 1. SWOT ANALYSIS

The SWOT analysis will be performed using a simple template shown in Figure 18 (already filled in with example information). It is important that participants have knowledge of the current situation regarding IT resources within the area of analysis (e.g., country, forest agency, forest enterprise) and a general understanding of what these resources are and what is their general purpose.

Each resource is recorded in a table and is assessed over two dimensions: internal and external factors. For each dimension, both positive and negative aspects should be identified.

FIGURE 18. **Results of an analysis of IT resources required to build the Forest Data Bank in Poland (performed before the actual system was built). Please see the list of abbreviations at the beginning of this document for explanation of the abbreviations used)**

	IT resource	Internal		External	
		Strengths	Weaknesses	Opportunities	Threats
1	Information System of State Forests	Well established. Operational in all forest inspectorates (FI). Covers 78% of Poland's forest area. Contains large amounts of up-to-date information. Information easy to retrieve in digital form.	Big system, not easy to modify and adjust to new requirements. Wrapped in bureaucratic management processes. Outdated data structures. No historical information stored (available in historical backups only). No automated data transfer possible, manual processes will have to be established.		Dependent on internal financing, governance changes (e.g., moving SFH under the state budget) could hinder keeping current scale of operation.
2	Qualified personnel at BULiGL	A competent team of IT specialists, often with forestry background. IT division has well-established project execution SOPs and tested modes of operation. ISO9001 in place.	Lack of experience in running large-scale hardware installations.	Acquiring a contract to build and maintain FDB will stabilize financial standing of the IT division.	Fast-growing salaries of IT specialists (especially in Warsaw area) which make BULiGL not competitive and causes increasing problems with recruitment.
3	IT infrastructure at BULiGL	BULiGL has its own high-quality server room to potentially host FDB infrastructure.	Insufficient capacity of Internet link. Investment required to provide high throughput and HA.	Internet link upgrade possibility exists on site.	
4	External contractors	BULiGL has business links with reliable contractors providing technical resources and assistance.		External contractors can provide know-how which is missing at BULiGL.	If improperly managed, using external contractors can develop critical dependency on them.

Practical note: The moderator presents a large table (on paper on the wall or on a whiteboard). Participants first insert the name of IT resource in the table rows. Once all important IT resources have been identified, participants assess each resource.

Care should be taken to distinguish internal and external factors. While this could be complex, users tend to understand the difference better over time. As a rule of thumb, anything that is a property of the resource should be considered internal, while all factors outside the resource location influences or organizations will be treated as external.

STEP 2. EVALUATION AND IDENTIFICATION OF POINTS OF INTERVENTION

The completion of the SWOT table should be followed by the identification of the crucial resources which require intervention to reach (or keep) the level sufficient from the point of view of FIS development, i.e. their condition, service level, quality, quantity, etc. There is no standardized routine for this, as it largely depends on local conditions and priorities. Figure 19 contains points of intervention identified as crucial for building the Forest Data Bank in Poland.

FIGURE 19. **Points of intervention identified as crucial for building the Forest Data Bank in Poland**

	IT resource	Intervention	Justification
1	IT infrastructure at BULiGL	Increase capacity of the Internet link to 100 Mb/s. Purchase and install an alternative link of at least 20 Mb/s from another provider with different technology. Configure high availability using the two links (possibly BGP).	Expected user load and data transfers (especially from GIS portal) are estimated to exceed the currently installed capacity. Current Internet link is not resistant to failure and HA is required for uninterrupted operation of the FBD portal.
2	Qualified personnel at BULiGL.	Provide stable funding and keep salary dynamics in line with the general market trend.	Ensure retaining stable experienced personnel with unique qualifications (IT + forest management).

It might turn out that some required interventions are so extensive and/or costly that this may jeopardize the viability of the entire FIS project. This would be important information for decision-makers to know at this point, as they might decide to re-evaluate their priorities and develop other solutions.

VII. DEFINING KEY PERFORMANCE INDICATORS

Objective of the exercise

The objective of this exercise is to design and develop specific, measurable, achievable, relevant and time-bound (SMART) key performance indicators (KPIs) in line with the related IT objectives to assess the FIS alignment against the objectives and the FIS efficiency criteria and thus, enable corrective action if needed.

Introduction

What are KPIs?

KPIs are a measurable value or metric to assess the performance and progress of an individual, team, department, or organization in achieving specific objectives or goals. KPIs are crucial in business and management as they provide a quantifiable way to track performance and ensure that efforts align with strategic objectives.

Key Performance Indicators are typically chosen based on their relevance to the organization's overall mission and objectives. They should be SMART to be effective in providing meaningful insights and driving improvements.

Examples of KPIs can vary widely depending on the nature of the organization and its goals. Some common KPIs in different functional contexts include:

- **Sales:** Revenue, sales growth rate, customer acquisition rate;
- **Customer Service:** Customer satisfaction score, average response time, customer retention rate;
- **Marketing:** Conversion rate, website traffic, cost per lead;
- **Finance:** Profit margin, return on investment (ROI), cash flow;
- **Human Resources:** Employee turnover rate, employee engagement score, training completion rate;
- **Manufacturing:** Production efficiency, defect rate, on-time delivery.

By regularly monitoring and analysing KPIs, businesses can make data-driven decisions, identify areas for improvement and focus their efforts on achieving success in line with their strategic objectives.

What is the purpose of defining KPIs in developing FIS?

It should be remembered that the business objectives for building/improving a forest information system and the corresponding IT-related objectives, have resulted from the analysis of what interventions could improve efficiency of the forest sector value chain (Phase 1). Tools are needed to assess the results of implementation of these strategies and, if necessary, to take corrective action. KPIs provide the necessary metrics for this purpose.

How FIS KPIs should be defined?

For each improvement strategy (linked to the business objectives), KPIs should be defined, as well as the means to assess them.

For example, for the business objective on "having control on illegal logging and deforestation" and the corresponding IT objective on "creating a timber-tracking system" the KPI could be: "the share of timber produced in the country which is tracked by the timber-tracking system", or: "the number of forest inspectorates which have implemented the timber-tracking system".

The KPIs must be SMART. They must provide the means for measuring the goals reached and the efficiency of critical processes linked to relevant business objectives. Potential KPIs could cover the following categories, with specifications:

Data accuracy: Measure the accuracy and reliability of the data stored and managed within the FIS via data validation rates of logic checks.

Data Timeliness: Measuring the timeliness of data updates and availability in the FIS.

System Performance: Assessing the performance and responsiveness of the FIS by including metrics on data and analysis processing.

System utilization and user satisfaction: Evaluating the extent to which the FIS is utilized (for example the number of administrative cases done using the FIS) and the level of satisfaction by stakeholders.

Decision-making support: Assessing the effectiveness of the FIS in supporting informed decision-making by tracking the number of decisions influenced by the system's outputs.

STEP 1. IDENTIFICATION OF KPIS

Identification of KPIs involves only 1 step. There are many approaches to defining KPIs. They depend on the context, organizational structure, objectives and level of the analysis/ assessment. An example of a KPI developed for the Forest Data Bank in Poland is shown in Figure 20.

FIGURE 20. **An example of the definition of a KPI measuring the extent to which the Forest Data Bank in Poland provides information on forests to the general public**

KPI name	Number of visits to the Forest Data Bank portal
Related objective	Provide information on forests to stakeholders and general public
KPI question	How popular is the Forest Data Bank?
How the KPI will be used	Internally to adjust required IT capacities. Externally for reporting to the contract provider (State Forests Holding)
Data collection method	Google Analytics
Target	20 000 visits per month
Assessment frequency	Once a month
Provider	BULiGL IT department
Assessment cost	2 man-hours a month

VIII. IDENTIFYING STRATEGIC ACTIONS

Objective of the exercise

The objective of this exercise is to prepare a strategic action plan for implementing a forest information system by considering key factors such as stakeholder needs, partnerships and technological infrastructure.

Introduction

Why do we need to identify strategic actions?

The previous exercises increased the understanding of stakeholders and institutions in the forest sector by mapping their relationships and dependencies. A value chain analysis (VCA) was carried out, which provided the foundation for identifying business-related objectives that should precede the FIS development. They were translated into IT-related objectives. Available critical IT resources were assessed and necessary points of intervention were identified to properly support the IT-related objectives. Finally, principal KPIs designed to monitor alignment of FIS development results to the objectives were formulated.

In the next step, specific strategic actions should be defined. They should lead to the elimination or at the least reduction of the gaps identified. Likewise, efficiency and quality of processes in the forest sector should be improved to support the overall objectives. These strategic actions will be the framework for the individual projects.

How should strategic actions be identified?

There is no one-fit-for-all approach as the local context, priorities and resources must be considered. One possible method is to begin with designing information systems around the forest management strategy, stemming from existing forest policy, the institutional landscape, organizational and financial capabilities and administrative capacity.

Identifying strategic actions for implementing a FIS involves careful consideration of key factors that contribute to its successful establishment and operation. After a successful stakeholder analysis and the mapping of institutions, understanding the specific requirements and priorities of various stakeholders is essential to formulating a strategic action plan that addresses their diverse needs effectively.

Establishing strong partnerships and collaboration with relevant stakeholders is vital for engaging with experts from the forest sector, academia, technology providers, as well as political actors, to foster a shared vision and encourage knowledge exchange. This can be done by involving different stakeholders in the planning process at an early stage to create a joint strategic action plan.

Ensuring the availability and accessibility of appropriate technology infrastructure is crucial. This involves evaluating existing technological capabilities, addressing potential limitations and implementing upgrades or new solutions as needed.

STEP 1. LIST STRATEGIC ACTIONS

This also involves only 1 step. Use all the materials prepared at earlier steps of this methodological framework to identify strategic actions to be undertaken. This process should be done with the involvement of decision makers from policy, business and IT areas. There are many possible ways to create a strategic action plan. It could be a simple text document outlining the steps and projects necessary to achieve the strategic goals. It is often beneficial to use a structured template, as it may help to include all necessary information. The degree of detail of such a template depends on the level of analysis. High-level strategic action plans are generally less detailed as their role is to outline only necessary actions. Low-level operational action plans are usually more detailed because they are later translated into specific projects with well-defined outcomes.

Figure 21 presents an example of a strategic action plan to achieve a strategic business object to improve the provision of forest-related information to the public in Poland. It is provided as a general example only, as each action plan may have a different level of detail and most probably different goals and associated business objectives.

FIGURE 21. **An example action plan to achieve a strategic business object to improve the provision of forest-related information to the public in Poland**

	Strategic action	End goal (SMART - specific, measurable, achievable, relevant and time-bound)	Steps to be followed (including individual projects)	Timeline and milestones	Resources needed
1	Build Forest Data Bank – a nation-wide information system providing comprehensive information on the Polish forests.	(S) Web portal + database system to hold all available information on Polish forests. (M) The system will contain and publish inventory data on all forests coming from the State Forests Information System, county administration, national parks and other managers of public forests. The system will publish information via a web portal. (A) The system will be built by BULiGL which has necessary expertise, know-how and resources, it will be financed by the State Forests Holding from the forest fund. (R) The system will fulfil obligations set in the Forest Act and will improve communication with the public on forest issues. (T) The project will start in 2010. The complete product with all planned data will be launched by April 2013.	Collect and standardize source data. Purchase and deploy IT hardware infrastructure and networking. Design and implement a database warehouse. Fill the data warehouse with data. Develop web portal application. System completed and tested.	Dec 2011 Dec 2010 Jun 2012 Dec 2012 Dec 2012 Apr 2013	2 backend developers (Java) 1 front-end developed 1 IT system administrator 2 GIS specialists 1 data specialist + tester 1 project manager External contract for hardware delivery and configuration Additional Internet link + BGP
2	Build general-purpose mobile app for easy access to information on forests.	(S) Mobile app (Android and iOS) that publishes information from FDB (M) Mobile app published in Google Play and App Store, publishing all thematic maps from FDB + GPS navigation and work offline (A) App will be built by BULiGL within a contract for FDB. (R) App will be a novel channel of communication of forest information to the public. (T) The app will be launched in April 2014.	Project to build the mobile FDB app	Apr 2014	1 mobile developed (Xamarin) Hardware for testing

REFERENCES

Bryson, J.M. 2018. Strategic planning for public and nonprofit organizations: A guide to strengthening and sustaining organizational achievement. John Wiley & Sons.

CMMI Product Team and others. 2006. CMMI for development, version 1.2. Software Engineering Institute, Pittsburgh USA.

Gërvalla, M., Preniqi, N., and Kopacek, P. 2018. IT Infrastructure Library (ITIL) framework approach to IT Governance. IFAC-PapersOnLine 51(30): 181–185. doi:10.1016/j.ifacol.2018.11.283.

Golder, B., and Gawler, M. 2005, October. Cross-Cutting Tool: Stakeholder Analysis. WWF. Available from https://awsassets.panda.org/downloads/1_1_stakeholder_analysis_11_01_05.pdf [accessed 9 May 2023].

Humphrey, J., and Navas-Alemán, L. 2010. Value Chains, Donor Interventions and Poverty Reduction: A Review of Donor Practice. IDS Research Reports 2010(63): 1–106. doi:10.1111/j.2040-0217.2010.00063_2.x.

Paulo, H. de S.B., Adriano, O.T., reacute, L.Z., Mozar, J. de B., and Joseacute, L.T. 2012. Implementation of information technology (IT) governance through IT strategic planning. Afr. J. Bus. Manage. 6(45): 11179–11189. doi:10.5897/AJBM11.2097.

Stein, C., and Barron, J. 2017. Mapping actors along value chains: integrating visual network research and participatory statistics into value chain analysis. International Water Management Institute (IWMI). CGIAR Research Program on Water, Land and Ecosystems (WLE). doi:10.5337/2017.216.

Storey, H.L., Agarwal, N., Cantera, J., Golden, A., Gallo, K., Herrick, T., Belizario, V., Kihara, J., Mwandawiro, C., Cadwallader, B., and De Los Santos, T. 2019. Formative research to inform development of a new diagnostic for soil-transmitted helminths: Going beyond the laboratory to ensure access to a needed product. PLoS Negl Trop Dis 13(5): e0007372. doi:10.1371/journal.pntd.0007372.

Suoheimo, M., Kays, R., Ogundaini, M., and Dammann, J. 2017. Participatory Tools for Human Development with the Youth. D3.3 New tools, methods and models for multi-stake-holder service and youth development working models. Available from https://ec.europa.eu/research/participants/documents/downloadPublic?documentIds=080166e5b77b5f3c&appId=PPGMS [accessed 6 June 2023].

Svadlenak-Gomez, K., Gerritsmann, H., Badura, M., and Walzer, C. 2014. Biodiversity Stakeholder Networks in the Alpine Space. greenAlps project/FIWI. doi:10.13140/2.1.4695.8401.

ANNEX: INFORMATION SYSTEM ANALYSIS RESOURCES

I. INSTITUTIONAL MAPPING

Stakeholder analysis worksheet

FIGURE 22. **A template for stakeholder analysis**

Stakeholder group name	Key-stakeholder	1. What is the stakeholder's primary interest in FIS?	2. To what degree is the stakeholder group relevant to developing FIS?	3. To what degree is the group involved in the policymaking process?	4. Would the stakeholder group support or oppose the FIS development?	5. Will the developing FIS benefit or harm the stakeholder group?	6. What alliances exist with other stakeholders?	7. What conflicts exist with other stakeholders?

FIGURE 23. **An example of the Stakeholder Analysis Worksheet based on the forest sector in Poland**

Note: this is a limited example of the forest sector in Poland.

Stakeholder group name	Stakeholder name	What is the stakeholder's primary interest in FIS?		To what degree is the stakeholder group relevant to developing FIS?		To what degree is the group involved in the policymaking process?		Would the stakeholder group support or oppose the FIS development?	Will the developing FIS benefit or harm the stakeholder group?		What alliances exist with other stakeholders?	What conflicts exist with other stakeholders?
		Score interest (1-5)	Describe how this stakeholder group relates to forest information processing, using or related policies	Score relevance (1-5)	1. What is the current role of the stakeholder in developing the FIS? 2. What is the potential role of the stakeholder in the future FIS development? 3. What is the potential contribution of the stakeholder to the FIS development?	Score involvement (1-5)	Explain, taking into account how this group has been involved in decision-making on policy both historically and presently.	Assign one of the following: Supporter, Moderate Supporter, Neutral, Moderate Opponent and Opponent	Answer this question with: Benefit, Harm, Neither, Mixed	Explain how	List and describe any collaboration between relevant stakeholders	Describe any conflicts between stakeholder groups
Government	Ministry of Climate and Environment	4	The ministry is primarily a consumer of information about natural resources (including forests). It needs accurate and timely information to be able to perform appropriate oversight over forests. The Minister is empowered to accept forest management plans for forests on state-owned lands. The ministry is also the contact point in most international reporting processes that relate to forestry (e.g., Forest Resources Assessment, Pan-European reporting etc.). The ministry is a primary source of information for the Parliament and the President of the Republic in the area of climate, nature and environment.	5	1. A few years ago the ministry commissioned an initial analysis of how a national set of criteria and indicators (C&I) for sustainable forest management could be developed. No follow-up so far. 2. The ministry could be an initiator of building/improving FIS for national forest statistics. The ministry also directly supervises national parks (23) and could initiate building an integrated FIS for the national parks in Poland. 3. The ministry could initiate and coordinate creating a national set of C&A for SFM as the framework for high-level reporting on the condition and trends in forests and forest industry. This potentially could lead to creating a whole system of reporting on C&As for SFM in Poland.	5	The ministry is the principal player in policymaking processes and participates in preparing legal regulations. The Minister approves the methodology and results of the National Forest Inventory. The ministry supervises the national parks and sets regulations for their work. It is also responsible for implementing legal obligations (both national and international) which might require information collecting, processing and reporting.	Moderate Supporter	Benefit	The ministry will receive information on forests which will be of better quality, timely, accurate and reliable. When needed, FIS may provide timely, accurate and reliable detailed data on particular problem areas.	Supervision over the General Director of State Forests and collaboration with the State Forests Holding over many issues related to forests. Supervision over the national parks.	The degree of collaboration of the State Forests Holding with the ministry depends on the actual political configuration of the ruling coalition. Conflicts with NGOs over how forests should be managed.

Stakeholder group name	Stakeholder name	What is the stakeholder's primary interest in FIS?		To what degree is the stakeholder group relevant to developing FIS?		To what degree is the group involved in the policymaking process?		Would the stakeholder group support or oppose the FIS development?	Will the developing FIS benefit or harm the stakeholder group?		What alliances exist with other stakeholders?	What conflicts exist with other stakeholders?
Special public bodies	General Directorate of State Forests (DGLP)	5	The General Directorate of State Forests needs to coordinate the work of 439 forest inspectorates. This requires intensive information exchange between FI and DGLP, between organizational units of DGLP and in some cases with external stakeholders. The FIS needs to be standardized within the whole organization. DGLP is also responsible for external reporting to the Ministry.	5	1. The State Forests Holding (SFH) is a separate public entity governed by dedicated legislation, with a significant degree of independence regarding how it is organized and how its activities are carried out. Since 1990s, SFH has been developing their own FIS as a fully-fledged ERM system, covering all business processes. DGLP coordinates and regulates the process of development of this FIS, as well as regulates its use. Moreover, the Forest Act obliges DGLP to carry out the National Forest Inventory (NFI) and run the Forest Data Bank. DGLP has a decisive role in setting directions and guidelines for both processes, as well as provide financial resources. DGLP maintains their own IT structures responsible for contracting external IT services, maintaining their data centers, maintaining crucial infrastructure, security and carrying out some software development projects. 2. DGLP does and will set directions on the development of their internal FIS, both business-related and technical. There are efforts to modernize technology in which their internal FIS is implemented. DGLP may initiate development of new functional modules of their FIS, covering new business processes (or the old ones which haven't been digitized yet). 3. As the SFH functions on the basis of self-financing, they do and will provide financial resources for development of their FIS and the Forest Data Bank.	4	DGLP has significant involvement in policymaking in Poland. All draft legal acts related to forests and forest land management are consulted with DGLP. DGLP officers actively participate in national and international cooperation on all levels. DGLP provides actionable information for the Ministry, lawmakers and other public agencies. They are invited to consulting bodies, task forces, discussion groups and other ad hoc structures often created to support decision-and policymaking.	Supporter	Benefit	Better information collection and processing could further improve efficiency of forest management by the SFH, improving its financial standing. Data-driven decision-making could improve quality of management. Technology could be a key to better communication with the public and improvement of the SFH's public image.	Supervised by the Minister of Climate and Environment. Collaboration with BULiGL (NFI, Forest Data Bank, forest management plannings, expertise) Collaboration with academia from forestry faculties.	Conflicts with NGOs over how forests should be managed.

Stakeholder group name	Stakeholder name	What is the stakeholder's primary interest in FIS?		To what degree is the stakeholder group relevant to developing FIS?		To what degree is the group involved in the policymaking process?		Would the stakeholder group support or oppose the FIS development?	Will the developing FIS benefit or harm the stakeholder group?		What alliances exist with other stakeholders?	What conflicts exist with other stakeholders?
State-owned enterprises	Bureau for Forest Management and Geodesy (BULiGL)	5	BULiGL is the biggest forest management planning organization in Poland (over 90% market share). Their primary role is to collect information on forests (forest inventory), process it in the course of forest management plan preparation (analysis and planning) and prepare the results in a standardized form ready for feeding into the FIS of the SFH. Forest information processing is intensively used in their core business (forest management planning). BULiGL is also legally assigned operator of the Forest Data Bank, having created its IT infrastructure, importing data from other sources, analysing and publishing in various communication channels.	5	1. BULiGL takes part in setting technical standards for forest information processing. It has recently participated in a consortium that has won a tender to develop a new technological platform for the SFH FIS and implement a forest management planning module. It is the operator of the Forest Data Bank. It has developed many IT solutions supporting forest management planning, forest inventory and forest data analysis. 2. BULiGL has significant potential in data processing and IT-related activities. At the same time, it has expertise and practical experience in forestry, so it could serve as a bridge between business and IT in forestry (especially forest management planning and communication with the public). 3. BULiGL has software engineering capacities so it could participate in developing FIS.	3	BULiGL is a traditional supplier of forest-related information for decision-makers. It publishes annual reports on the extent and condition of forest resources in Poland. In the past it took part in providing expert assessments and forecasts of various parameters in the Polish forestry. BULiGL is the executioner of the NFI in Poland which provides information on forests to the national statistics and to the general public.	Supporter	Benefit	Developing FIS will improve data accessibility and reliability which is important for the purposes of forest management planning and running the Forest Data Bank. Formal requirements regarding data structures in forest management planning can improve BULiGL's competitive advantage over smaller private forest management planning companies.	Collaboration with DGLP, some private technology providers and some academia	
Special public bodies	Forest managers (State Forests)	5	Most of business processes in the State Forests Holding have been digitized and forest managers are the principal users of the SFH internal FIS.	5	1. Forest managers are the principal users of the SFH internal FIS, utilizing information in it, providing updates and using it for their daily activities as well as for reporting and resource management. 2. User of the internal FIS in the SFH. 3. Providing requirements, suggesting new or modified use cases, providing and testing use case scenarios, test new/updated functions.	1	Forest inspectorates' role is to manage forests, make operational decisions within the boundaries of law and internal regulations and implement directives coming from the higher levels of the SFH structure.	Moderate Supporter	Benefit	Automation could further reduce human workload and improve working conditions. Development of FIS could result in more ergonomic and user-friendly solutions. Increased use of mobile devices covering more business processes and use cases could further improve working conditions and efficiency. IoT sensors, cameras etc. could further improve efficiency of forest fire protection, better prevent timber theft and improve public security.	Supervised by DGLP, collaboration with BULiGL, good relationships with some private technology providers	

Stakeholder group name	Stakeholder name	What is the stakeholder's primary interest in FIS?		To what degree is the stakeholder group relevant to developing FIS?		To what degree is the group involved in the policymaking process?		Would the stakeholder group support or oppose the FIS development?	Will the developing FIS benefit or harm the stakeholder group?		What alliances exist with other stakeholders?	What conflicts exist with other stakeholders?
Local government	County offices	3	Acquire and use software solutions improving efficiency and quality of the supervision of private forests. Tools to provide data on forests required by law or cooperation with the SFH.	3	1. Customers requiring solutions to support the supervision of private forests. 2. User of software solutions supporting the supervision of private forests. 3. Providing requirements, suggesting new or modified use cases, providing and testing use case scenarios, test new/updated functions.	1	County office's role is to supervise management of private forests withing the boundaries of law.	Moderate Supporter	Benefit	Automation could further reduce human workload and improve working conditions. Development of FIS could result in more ergonomic and user-friendly solutions, as well as provide non-technical users with tools to better understand the situation (e.g., GIS) and to exchange data with other organizations/units.		
Private persons or legal entities	Private forest managers	1	Most of private forests and very small and managed extensively. No significant interest in FIS in this stakeholder group as it wouldn't be economically viable.	1	1. Most of private forests and very small and managed extensively. No significant interest in FIS in this stakeholder group as it wouldn't be economically viable. 2. User of a forest management support system for private owners. User of a timber trade platform. 3. None.	1	Not involved at all.	Neutral	Benefit	If new legal obligations were introduced to provide data on forest management in structured digital format, an FIS for private owners could give them tools to easily do that. Private forest owners could potentially benefit from creating a timber trade platform where they could easily sell their timber.		
Special public bodies	National parks	5	Acquire and use software solutions for comprehensive management of forests and other lands in national parks.	5	1. Forest ecosystems are an important part of natural resources in national parks in Poland. National parks want to have IT tools to support management. Some of them use solutions available on the market. 2. Participation in future FIS development programs. Using the FIS. 3. Lobbying towards standardization of IT tools and data formats. Participation in FIS development programs (some resources may be available only to public agencies and entities, e.g., EU developmental programs). Providing requirements, suggesting new or modified use cases, providing and testing use case scenarios, test new/updated functions.	2	Not involved at all at the moment but some influence is exerted indirectly via the Ministry and horizontal collaboration.	Supporter	Benefit	No FIS covering all business processes in national parks are developed at the moment. If a unified FIS were developed, its cost per user would be much lower. The future FIS would create a platform for national park officers to integrate all data, including inventory and research data. Automation could further reduce human workload and improve working conditions. Development of FIS could result in more ergonomic and user-friendly solutions. Increased use of mobile devices covering more business processes and use cases could further improve working conditions and efficiency. IoT sensors, cameras etc. could further improve efficiency of forest fire protection, better prevent timber theft and improve public security.	Each national park is supervised by the Minister. Horizontal collaboration between national parks. Often collaboration with BULiGL and private technology providers.	

Stakeholder group name	Stakeholder name	What is the stakeholder's primary interest in FIS?		To what degree is the stakeholder group relevant to developing FIS?		To what degree is the group involved in the policymaking process?		Would the stakeholder group support or oppose the FIS development?	Will the developing FIS benefit or harm the stakeholder group?		What alliances exist with other stakeholders?	What conflicts exist with other stakeholders?
Government agencies	Forest managers (other state-owned forests)	3	Acquire and use software solutions for comprehensive management of forests and other lands in other state-owned forests. There are not too much of them.	5	1. Nothing really. They use old SFH-derived FIS without much development. 2. Participation in future FIS development programs. Using the FIS. 3. Providing requirements, suggesting new or modified use cases, providing and testing use case scenarios, test new/updated functions.	1	Not involved at all.	Moderate Supporter	Benefit	No FIS covering all business processes in these organizations are developed at the moment. If a unified FIS were developed, its cost per user would be much lower. The future FIS would create a platform for officers to integrate all data, including inventory and research data. Automation could further reduce human workload and improve working conditions. Development of FIS could result in more ergonomic and user -friendly solutions. Increased use of mobile devices covering more business processes and use cases could further improve working conditions and efficiency. IoT sensors, cameras etc. could further improve efficiency of forest fire protection, better prevent timber theft and improve public security.	Sometimes collaboration with BULiGL	
Forest managers (municipal)		4	Acquire and use software solutions for comprehensive management of municipal forests.	4	1. Nothing really. They use old SFH-derived FIS without much development. 2. Participation in future FIS development programs. Using the FIS. 3. Providing requirements, suggesting new or modified use cases, providing and testing use case scenarios, test new/updated functions.	1	Not involved at all.	Moderate Supporter	Benefit	No FIS covering all business processes in these organizations are developed at the moment. If a unified FIS were developed, its cost per user would be much lower. The future FIS would create a platform for officers to integrate all data, including inventory and research data. Automation could further reduce human workload and improve working conditions. Development of FIS could result in more ergonomic and user -friendly solutions. Increased use of mobile devices covering more business processes and use cases could further improve working conditions and efficiency. IoT sensors, cameras etc. could further improve efficiency of forest fire protection, better prevent timber theft and improve public security.		Conflicts with NGOs over how forests should be managed.

Stakeholder group name	Stakeholder name	What is the stakeholder's primary interest in FIS?		To what degree is the stakeholder group relevant to developing FIS?		To what degree is the group involved in the policymaking process?		Would the stakeholder group support or oppose the FIS development?	Will the developing FIS benefit or harm the stakeholder group?		What alliances exist with other stakeholders?	What conflicts exist with other stakeholders?
Private persons or legal entities	Private solution providers	5	Develop and commercialize software solutions for forest management by various stakeholders.	5	1. There are several specialized IT companies developing IT solutions for forestry and nature management. Some of them are also consulted by their customers (especially SFH) for advice on technology, formats, hardware etc. 2. Advising on technology, solutions etc. Providing solutions and services. 3. Technical and business expertise. Providing ready solutions (installable or SaaS).	2	Consulted sometimes regarding data formats and technology.	Moderate Supporter	Mixed	Benefits: opportunities for large-scale projects connected with the FIS; opportunities for auxiliary tools, complementing the FIS. Harms: potentially closing market for their existing solutions if the FIS functionalities will overlap.	Some of them collaborate closely with the State Forests Holding.	
Non-governmental organizations	Ecoactivist NGOs	5	Easy access to detailed data on forests	1	1. None. 2. None. 3. None.	3	Soft-power influence on decision-makers	Supporter	Benefit	New/better/easier to access sources of data on forests.	Horizontal alliances between activists and organizations. Alliances with parts of academia (especially faculties of biology at universities).	Conflicts with most forest managers (excluding national parks) over many aspects of forest management.
Academia		5	Easy access to detailed data on forests	3	1. None. 2. Advisory activities. 3. Expert knowledge on forest management and planning.	2	Members of the academia are often members of task groups and advisory bodies, sometimes they are nominated for political functions.	Supporter	Benefit	New/better/easier to access sources of data on forests.	Academia from faculties of forestry provide expertise to the State Forest Holding. Academia from faculties of biology usually form alliances with environmental NGOs.	
Special public bodies	Statistics Poland	2	Easy access to detailed data on forests	2	1. None. 2. Consulting solutions, data formats etc. 3. Providing expertise on data required by the public statistics and their exchange formats.	3	Involved in the areas related to national statistics.	Moderate Supporter	Benefit	New/better/easier to access sources of data on forests.		

FIGURE 24. **A template for listing IT-related objectives**

	Business objective	IT-related objective	Description and justification (how this solution can contribute to achieving of the business objective)
1			

FIGURE 25. **A template to record the results of a SWOT analysis of IT resources**

	IT resource	Internal		External	
		Strengths	Weaknesses	Opportunities	Threats
1					

FIGURE 26. **A template to list required points of intervention regarding crucial IT resources**

	IT resource	Intervention	Justification
1			

II. DEFINING KEY PERFORMANCE INDICATOR

FIGURE 27. **A simple description template for defining a KPI**

	KPI name	
	IT-related objective	
	KPI question	
	How the KPI will be used	
	Data collection method	
	Target	
	Assessment frequency	
	Provider	
	Assessment cost	

III. IDENTIFYING STRATEGIC ACTIONS

FIGURE 28. **A simple template to define strategic actions**

	Strategic action	End goal (SMART)	Steps to be followed (including individual projects)	Timeline and milestones	Resources needed
1					

SOME FACTS ABOUT COFFI AND EFC

Some facts about the Committee on Forests and the Forest Industry

The UNECE Committee on Forests and the Forest Industry (COFFI) is a principal subsidiary body of the UNECE based in Geneva. It constitutes a forum for cooperation and consultation between member countries on forestry, the forest industry and forest product matters. All countries of Europe and the EECCA, as well as the United States, Canada and Israel, are members of the UNECE and participate in its work.

The UNECE Committee on Forests and the Forest Industry shall, within the context of sustainable development, provide member countries with the information and services needed for policymaking and decision-making with regard to their forest and forest industry sectors, including the trade and use of forest products and, where appropriate, it will formulate recommendations addressed to member governments and interested organizations. To this end, it shall:

1. with the active participation of member countries, undertake short-, medium- and long-term analyses of developments in and having an impact on, the sector, including those developments offering possibilities for facilitating international trade and for enhancing the protection of the environment;
2. in support of these analyses, collect, store and disseminate statistics relating to the sector and carry out activities to improve their quality and comparability;
3. provide a framework for cooperation, for example by organizing seminars, workshops and ad hoc meetings and setting up time-limited ad hoc groups, for the exchange of economic, environmental and technical information between governments and other institutions of member countries required for the development and implementation of policies leading to the sustainable development of the sector and the protection of the environment in their respective countries;
4. carry out ww identified by the UNECE or the Committee on Forests and the Forest Industry as being of priority, including the facilitation of subregional cooperation and activities in support of the economies in transition of central and eastern Europe and of the countries of the region that are developing from an economic perspective; and
5. keep under review its structure and priorities and cooperate with other international and intergovernmental organizations active in the sector, particularly FAO and its European Forestry Commission and the International Labour Organization, in order to ensure complementarity and avoid duplication, thereby optimizing the use of resources.

Some facts about the European Forestry Commission

The European Forestry Commission (EFC), which was created in 1947, is one of six regional forestry commissions established by FAO to provide a policy and technical forum for countries to discuss and address forest issues on a regional basis.

The purpose of the EFC is to advise on the formulation of forest policies and to review and coordinate their implementation at the regional level; exchange information; advise on suitable practices and actions to address technical and economic problems (generally through special subsidiary bodies); and make appropriate recommendations in relation to the foregoing. The EFC meets every two years and its official languages are English, French and Spanish.

The EFC has a number of associated subsidiary bodies, including the Working Party on the Management of Mountain Watersheds; the UNECE/FAO Working Party on Forest Statistics, Economics and Management; and seven UNECE/FAO Teams of Specialists. The Committee on Mediterranean Forestry Issues (Silva Mediterranea) informs the EFC.

FAO encourages the wide participation of government officials from forestry and other sectors as well as representatives of international, regional and subregional organizations that deal with forest-related issues in the region, including non-governmental organizations and the private sector. Accordingly, the EFC is open to all members and associate members whose territories are situated wholly or in part in the European Region or who are responsible for the international relations of any non-self-governing territory in that region. Membership comprises such eligible member nations as have notified the Director-General of their desire to be considered as members.

The EFC is one of the technical commissions serving the FAO Regional Office for Europe and Central Asia (REU) and the EFC Secretary is based in Geneva. EFC work is regulated by its Rules of Procedures, which were adopted by the FAO Conference in 1961 and amended at the Eighteenth Session of the EFC in 1977.

More information about the work of the EFC and COFFI may be obtained by contacting:

UNECE/FAO Forestry and Timber Section
Forests, Land and Housing Division
United Nations Economic Commission for Europe/
Food and Agriculture Organization of the United Nations
Palais des Nations
CH-1211 Geneva 10, Switzerland

info.ECE-FAOforests@un.org
www.unece.org/forests

UNECE/FAO PUBLICATIONS

Geneva Timber and Forest Study Papers

Reporting on Forest Damages and Disturbances in the UNECE Region	ECE/TIM/SP/57
Developing National Forest Information Systems: A Practical Guide	ECE/TIM/SP/56
Forest Products Annual Market Review 2022-2023	ECE/TIM/SP/55
Forest Products Annual Market Review 2021-2022	ECE/TIM/SP/54
Reporting on Forests and Sustainable Forest Management in the Caucasus and Central Asia	ECE/TIM/SP/53
Forest Products Annual Market Review 2020-2021	ECE/TIM/SP/52
Forest Sector Outlook Study, 2020-2040	ECE/TIM/SP/51
Forest Products Annual Market Review 2019-2020	ECE/TIM/SP/50
Circularity Concepts in Forest-Based Industries	ECE/TIM/SP/49
Forest Products Annual Market Review 2018-2019	ECE/TIM/SP/48
State of Forests of the Caucasus and Central Asia	ECE/TIM/SP/47
Forest Products Annual Market Review 2017-2018	ECE/TIM/SP/46
Forests and Water	ECE/TIM/SP/44
Forest Ownership in the ECE Region	ECE/TIM/SP/43
Wood Energy in the ECE Region	ECE/TIM/SP/42
Forest Products Annual Market Review 2016-2017	ECE/TIM/SP/41
Forest Products Annual Market Review 2015-2016	ECE/TIM/SP/40
Forest Products Annual Market Review 2014-2015	ECE/TIM/SP/39

Promoting Sustainable Building Materials and the Implications on the Use of Wood in Buildings	ECE/TIM/SP/38
Forests in the ECE Region: Trends and Challenges in Achieving the Global Objectives on Forests	ECE/TIM/SP/37
Forest Products Annual Market Review 2013-2014	ECE/TIM/SP/36
Rovaniemi Action Plan for the Forest Sector in a Green Economy	ECE/TIM/SP/35
The Value of Forests: Payments for Ecosystem Services in a Green Economy	ECE/TIM/SP/34
Forest Products Annual Market Review 2012-2013	ECE/TIM/SP/33
The Lviv Forum on Forests in a Green Economy	ECE/TIM/SP/32
Forests and Economic Development: A Driver for the Green Economy in the ECE Region	ECE/TIM/SP/31
Forest Products Annual Market Review 2011-2012	ECE/TIM/SP/30
The North American Forest Sector Outlook Study 2006-2030	ECE/TIM/SP/29
European Forest Sector Outlook Study 2010-2030	ECE/TIM/SP/28
Forest Products Annual Market Review 2010-2011	ECE/TIM/SP/27
Private Forest Ownership in Europe	ECE/TIM/SP/26
Forest Products Annual Market Review 2009-2010	ECE/TIM/SP/25
Forest Products Annual Market Review 2008-2009	ECE/TIM/SP/24
Forest Products Annual Market Review 2007-2008	ECE/TIM/SP/23
Forest Products Annual Market Review 2006-2007	ECE/TIM/SP/22
Forest Products Annual Market Review, 2005-2006	ECE/TIM/SP/21
European Forest Sector Outlook Study: 1960 – 2000 – 2020, Main Report	ECE/TIM/SP/20
Forest policies and institutions of Europe, 1998-2000	ECE/TIM/SP/19
Forest and Forest Products Country Profile: Russian Federation	ECE/TIM/SP/18

(Country profiles also exist on Albania, Armenia, Belarus, Bulgaria, former Czech and Slovak Federal Republic, Estonia, Georgia, Hungary, Lithuania, Poland, Romania, Republic of Moldova, Slovenia and Ukraine) Forest resources of Europe, CIS, North America, Australia, Japan and New Zealand

Note: Additional market-related information is available in electronic format at www.unece.org/forests.

Geneva Timber and Forest Discussion Papers (original language only)

Circularity Concepts in Pulp and Paper Industry	ECE/TIM/DP/97
Sustainable and Circular Bioeconomy in Forest-Based Industries	ECE/TIM/DP/96
Circularity Concepts in Wood Construction	ECE/TIM/DP/95
Forecast of the Committee on Forests and the Forest Industry Forest Products Production and Trade 2022-2023	ECE/TIM/DP/94
The Outlook for the UNECE Forest Sector in a Changing Climate	ECE/TIM/DP/93
Structural Changes in the Forest Sector and Their Long-Term Consequences for the Forest Sector	ECE/TIM/DP/92
Catalogue of Wood Waste Classifications in the UNECE Region	ECE/TIM/DP/91
Detailed methodology for the Preparation of the Forest Sector Outlook Study 2020-2040	ECE/TIM/DP/90
COVID-19 Impacts on the Forest Sector in Countries in the Western Balkans	ECE/TIM/DP/89
COVID-19 Impacts on the Forest Sector in Countries Eastern Europe, Caucasus and Central Asia	ECE/TIM/DP/88
Forest Landscape Restoration in Eastern and South-East Europe	ECE/TIM/DP/87
Forecast of the Committee on Forests and the Forest Industry: Forest Products Production and Trade 2020-2022	ECE/TIM/DP/86
Overview of the State of Forests and Forest Management in Uzbekistan	ECE/TIM/DP/85
Overview of the State of Forests and Forest Management in Turkmenistan	ECE/TIM/DP/84
Overview of the State of Forests and Forest Management in Tajikistan	ECE/TIM/DP/83
Overview of the State of Forests and Forest Management in Kyrgyzstan	ECE/TIM/DP/82
Overview of the State of Forests and Forest Management in Kazakhstan	ECE/TIM/DP/81
Overview of the State of Forests and Forest Management in Georgia	ECE/TIM/DP/80
Overview of the State of Forests and Forest Management in Azerbaijan	ECE/TIM/DP/79
Overview of the State of Forests and Forest Management in Armenia	ECE/TIM/DP/78
Guidelines on the Promotion of Green Jobs in Forestry	ECE/TIM/DP/77
Forest Sector Workforce in the UNECE Region	ECE/TIM/DP/76
Rovaniemi Action Plan for the Forest Sector in a Green Economy: Mid-term Review	ECE/TIM/DP/75
Trends in Green Jobs in the Forest Sector in the UNECE Region	ECE/TIM/DP/74
Guidelines for the Development of a Criteria and Indicator Set for Sustainable Forest Management	ECE/TIM/DP/73

Forest Landscape Restoration in the Caucasus and Central Asia	ECE/TIM/DP/72
Green Jobs in the Forest Sector	ECE/TIM/DP/71
Measuring the Value of Forests in a Green Economy	ECE/TIM/DP/70
Forecast of the Committee on Forests and the Forest Industry: Forest Products Production and Trade 2016-2018	ECE/TIM/DP/69
Forecast of the Committee on Forests and the Forest Industry: Forest Products Production and Trade 2015-2017	ECE/TIM/DP/68
ECE Committee on Forests and the Forest Industry and European Forestry Commission: 70 Years Working in the Service of Forests and People	ECE/TIM/DP/67
Pilot Project on the System for the Evaluation of the Management of Forests (SEMAFOR)	ECE/TIM/DP/66
Comparative Assessment of Wood Biomass for Energy in Europe	ECE/TIM/DP/65
Forecast of the Committee on Forests and the Forest Industry: Forest Products Production and Trade 2014-2016	ECE/TIM/DP/64
Forecast of the Committee on Forests and the Forest Industry: Forest Products Production and Trade 2013-2015	ECE/TIM/DP/63
Competitiveness of the European Forest Sector	ECE/TIM/DP/62
Forecast of the Committee on Forests and the Forest Industry: Forest Products Production and Trade 2012-2014	ECE/TIM/DP/61
Forecast of the Committee on Forests and the Forest Industry: Forest Products Production and Trade 2011-2013	ECE/TIM/DP/60
Econometric Modelling and Projections of Wood Products Demand, Supply and Trade in Europe	ECE/TIM/DP/59
Swedish Forest Sector Outlook Study	ECE/TIM/DP/58
The Importance of China's Forest Products Markets to the UNECE Region	ECE/TIM/DP/57
Good Practice Guidance on Sustainable Mobilisation of Wood: Proceedings from the Grenoble Workshop	ECE/TIM/DP/56
Harvested Wood Products in the Context of Climate Change Policies: Workshop Proceedings - 2008	ECE/TIM/DP/55
The Forest Sector in the Green Economy	ECE/TIM/DP/54
National Wood Resources Balances: Workshop Proceedings	ECE/TIM/DP/53
Potential Wood Supply in Europe	ECE/TIM/DP/52
Wood Availability and Demand in Europe	ECE/TIM/DP/51
Forest Products Conversion Factors for the UNECE Region	ECE/TIM/DP/49

Mobilizing Wood Resources: Can Europe's Forests Satisfy the Increasing Demand for Raw Material and Energy Under Sustainable Forest Management? Workshop Proceedings - January 2007	ECE/TIM/DP/48
European Forest Sector Outlook Study: Trends 2000-2005 Compared to the EFSOS Scenarios	ECE/TIM/DP/47
Forest and Forest Products Country Profile; Tajikistan	ECE/TIM/DP/46
Forest and Forest Products Country Profile: Uzbekistan	ECE/TIM/DP/45
Forest Certification – Do Governments Have a Role?	ECE/TIM/DP/44
International Forest Sector Institutions and Policy Instruments for Europe: A Source Book	ECE/TIM/DP/43
Forests, Wood and Energy: Policy Interactions	ECE/TIM/DP/42
Outlook for the Development of European Forest Resources	ECE/TIM/DP/41
Forest and Forest Products Country Profile: Serbia and Montenegro	ECE/TIM/DP/40
Forest Certification Update for the UNECE Region, 2003	ECE/TIM/DP/39
Forest and Forest Products Country Profile: Republic of Bulgaria	ECE/TIM/DP/38
Forest Legislation in Europe: How 23 Countries Approach the Obligation to Reforest, Public Access and Use of Non-Wood Forest Products	ECE/TIM/DP/37
Value-Added Wood Products Markets, 2001-2003	ECE/TIM/DP/36
Trends in the Tropical Timber Trade, 2002-2003	ECE/TIM/DP/35
Biological Diversity, Tree Species Composition and Environmental Protection in the Regional FRA-2000	ECE/TIM/DP/33
Forestry and Forest Products Country Profile: Ukraine	ECE/TIM/DP/32
The Development of European Forest Resources, 1950 To 2000: a Better Information Base	ECE/TIM/DP/31
Modelling and Projections of Forest Products Demand, Supply and Trade in Europe	ECE/TIM/DP/30
Employment Trends and Prospects in the European Forest Sector	ECE/TIM/DP/29
Forestry Cooperation with Countries in Transition	ECE/TIM/DP/28
Russian Federation Forest Sector Outlook Study	ECE/TIM/DP/27
Forest and Forest Products Country Profile: Georgia	ECE/TIM/DP/26
Forest Certification Update for the UNECE Region, Summer 2002	ECE/TIM/DP/25
Forecasts of Economic Growth in OECD and Central and Eastern European Countries for the Period 2000-2040	ECE/TIM/DP/24
Forest Certification Update for the UNECE Region, Summer 2001	ECE/TIM/DP/23

Structural, Compositional and Functional Aspects of Forest Biodiversity in Europe	ECE/TIM/DP/22
Markets for Secondary Processed Wood Products, 1990-2000	ECE/TIM/DP/21
Forest Certification Update for the UNECE Region, Summer 2000	ECE/TIM/DP/20
Trade and Environment Issues in the Forest and Forest Products Sector	ECE/TIM/DP/19
Multiple Use Forestry	ECE/TIM/DP/18
Forest Certification Update for the UNECE Region, Summer 1999	ECE/TIM/DP/17
A Summary of "The Competitive Climate for Wood Products and Paper Packaging: the Factors Causing Substitution with Emphasis on Environmental Promotions"	ECE/TIM/DP/16
Recycling, Energy and Market Interactions	ECE/TIM/DP/15
The Status of Forest Certification in the UNECE Region	ECE/TIM/DP/14
The Role of Women on Forest Properties in Haute-Savoie (France): Initial Research	ECE/TIM/DP/13
Interim Report on the Implementation of Resolution H3 of the Helsinki Ministerial	
Conference on the Protection of Forests in Europe (Results of the second enquiry)	ECE/TIM/DP/12
Manual on Acute Forest Damage	ECE/TIM/DP/7

Many of the above publications can be downloaded free of charge at:

www.unece.org/forests

Publications and subscriptions are available through United Nations Publications Offices as follows:

Sales and Marketing Section, Room DC2-853
United Nations
2 United Nations Plaza
New York, NY 10017
United States of America
Fax: + 1 212 963 3489
E-mail: publications@un.org
Web site: https://shop.un.org/